Simple Solutions.

Minutes a Day-Mastery for a Lifetime!

Algebra I
Part B

Mathematics

Nancy L. McGraw

Bright Ideas Press, LLC
Cleveland, Ohio

Simple Solutions Algebra I Part B

Printed in the United States of America

ISBN-13: 978-1-934210-20-8
ISBN-10: 1-934210-20-x

Cover Design: Dan Mazzola
Editor: Kimberly A. Dambrogio

Welcome to Simple Solutions

Note to the Student:

This workbook will give you the opportunity to practice skills you have learned in previous grades. By practicing these skills each day, you will gain confidence in your math ability.

Using this workbook will help you understand math concepts easier and for many of you, it will give you a more positive attitude toward math in general.

In order for this program to help you be successful, it is extremely important that you do a lesson every day. It is also important that you check your answers and ask your teacher for help with the problems you didn't understand or that you did incorrectly.

If you put forth the effort, Simple Solutions will change your opinion about math forever.

Lesson #1

1. Solve for x. $5x-3=2x+12$

2. $40 \div 8+3 \cdot 7-12 \div 4=?$

3. Write $\frac{2}{5}$ as a decimal and a percent.

4. Solve for x. $-2x-10 \geq 4$

5. $8\frac{1}{5}-6\frac{4}{5}=?$

6. The slope of a horizontal line is always __________.

7. $6.5 \div 4=?$

8. Solve for x. $6x=90$

9. What number is 70% of 80?

10. $70+(-48)=?$

11. Find the slope of a line passing through the points (4, 2) and (–2, –1).

12. How many feet are in 2 miles?

13. $-4<x+1<7$ Graph the solution on a number line.

14. Find $\frac{3}{7}$ of 28.

15. Solve for a. $\frac{1}{5}a+8=15$

16. Evaluate $\frac{xy}{4}+9$ if $x=2$ and $y=8$.

17. Simplify. $\frac{14x^3y^3z^2}{21xy^2z}$

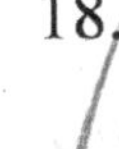

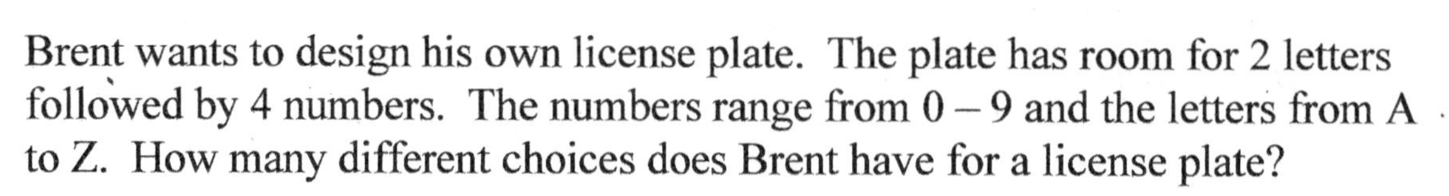

18. Brent wants to design his own license plate. The plate has room for 2 letters followed by 4 numbers. The numbers range from 0 – 9 and the letters from A to Z. How many different choices does Brent have for a license plate?

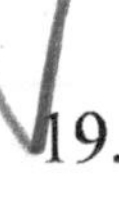

19. Solve for a. $\frac{a}{8}=\frac{21}{12}$

20. $6.35-4.9=?$

1.	2.	3.	4.
$X=5$	23	0.4	$x \leq -20$
5.	**6.**	**7.**	**8.**
$2\frac{-3}{0}$	1/2	1.625	$X=15$
9.	**10.**	**11.**	**12.**
0.56	22	?	200 ft
13.	**14.**	**15.**	**16.**
?	?	35	13
17.	**18.**	**19.**	**20.**
?	?	14	1.45

Lesson #2

1. Solve for x. $\frac{1}{7}x = 16$

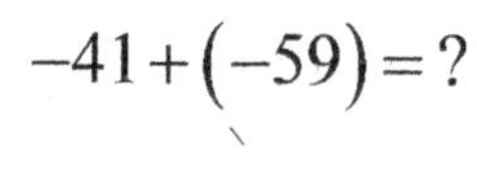

2. $-41 + (-59) = ?$

3. Find the GCF of $12a^2b^4c^3$ and $18ab^3c^5$.

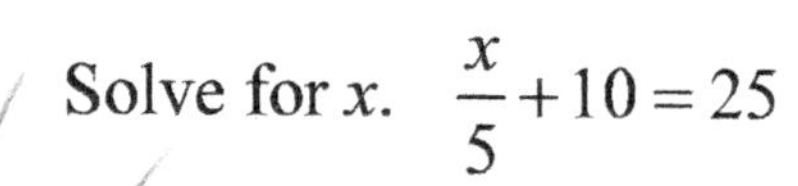

4. Solve for x. $\frac{x}{5} + 10 = 25$

5. Write $\frac{7}{20}$ as a decimal and as a percent.

6. $6\frac{1}{5} + 9\frac{1}{3} = ?$

7. Solve for x. $3x + 6 = 8x - 34$

8. Kim, who plays piano, has a finger span on her left hand of $8\frac{5}{16}$ inches from her thumb to her fifth finger. The span of her right hand is $8\frac{3}{4}$ inches. How much further can Kim stretch her right hand than her left?

9. $\sqrt{100} + 2^3 = ?$

10. Solve for x. $x - 4 \leq 2$

11. Find the missing measurement, x.

$x°$

$66°$

12. $3.5 \times 2.4 = ?$

13. Find the percent of change from $125 to $160.

14. What is the P(H, T, T, H) on 4 flips of a coin?

15. Solve for b. $b + 35 = -76$

16. $10 + 5[3(4 + 2) - 5] = ?$

17. Find the slope of a line passing through points (8, 3) and (– 4, 3).

18. Evaluate $7x - y$ if $x = 3$ and $y = 6$.

19. $60{,}000 - 27{,}314 = ?$

20. Solve for x. $-8 < 2x + 10 < 14$

1.	2.	3.	4.
$x = 112$	-100	$30a^2b^7c^8$	$x = 75$
5.	**6.**	**7.**	**8.**
0.35	$15\,2/8$	$8 = x$	133
9.	**10.**	**11.**	**12.**
10.39	$x \leq 6$	24	8.4
13.	**14.**	**15.**	**16.**
$x = 78.125$	?	$b = -111$	60
17.	**18.**	**19.**	**20.**
0	15	32,686	$x = -1$

Lesson #3

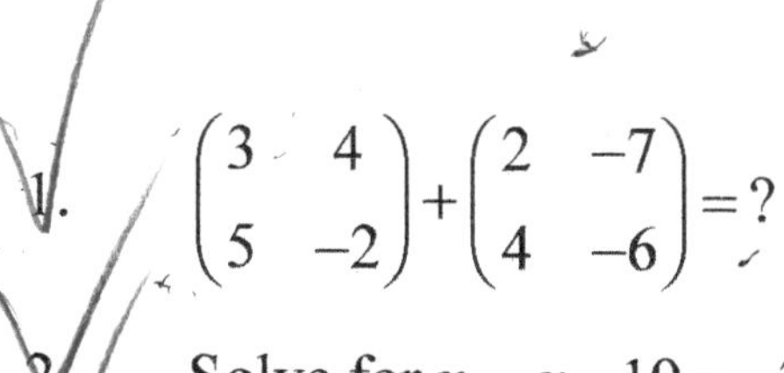

1. $\begin{pmatrix} 3 & 4 \\ 5 & -2 \end{pmatrix} + \begin{pmatrix} 2 & -7 \\ 4 & -6 \end{pmatrix} = ?$

2. Solve for x. $x - 19 = -31$
3. $152 + (-78) = ?$
4. Which is greater, $\frac{1}{5}$ or 26%?
5. Mrs. Jackson divided the 48-foot-wide stage into sections $4\frac{4}{5}$ feet long in order to prepare for the school program. How many sections were there?
6. Solve for x. $\frac{x}{18} = \frac{45}{15}$
7. How many ounces are in 4 pounds?
8. $\frac{4}{7} \times \frac{14}{20} = ?$
9. Solve for a. $3a - 15 = 15$
10. $-6(-3)(-2) = ?$
11. Find the surface area of the prism.

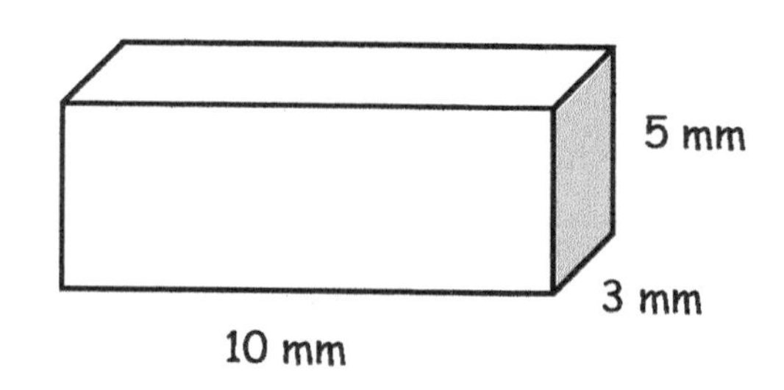

12. $81 \div 9 + 4 \cdot 2 + 3 \cdot 3 - 1 = ?$
13. How many centimeters are in 6 meters?
14. Solve for a. $5a - 7 = 12a - 7$
15. Find the values for y in the equation $y = 3x - 4$ when $x = \{0, -4, 2\}$.
16. $4 + a > 3$ or $6a < -30$ Solve the inequality for a and graph.
17. Write an algebraic phrase for *sixteen divided by a number less ten.*
18. Solve for x. $\frac{x}{9} + 6 = 12$
19. Simplify. $3(4x - 2y + 5) + 2(5x + 3y - 4)$
20. 60% of 80 is what number?

1. 3	2. $x=-12$	3. 74	4. 26%
5. 10	6. $x=54$	7. ?.	8. 0.4
9. $a=10$	10. −36	11. 18	12. 25
13. ?	14. $0=a$	15. $y=-16$	16. $a<-5$
17. 1.6	18. 54	19. ?.	20. $n=48$

Lesson #4

1. The ratio of chairs to tables in a restaurant is 12 to 7. If there are 84 chairs, how many tables are in the restaurant?

2. Find the slope of a line passing through points (–8, 0) and (1, 5).

3. Write 85% as a decimal and as a reduced fraction.

4. $-18+(-16)+9=?$

5. $\begin{pmatrix} 4 & -6 \\ 9 & 0 \end{pmatrix} - \begin{pmatrix} -8 & 3 \\ 0 & -4 \end{pmatrix} = ?$

6. $60-8(5-2)+7\cdot 3=?$

7. How many quarts are in 9 gallons?

8. What is the value of n? $n-26=-50$

9. Solve for x. $7x-9=19$

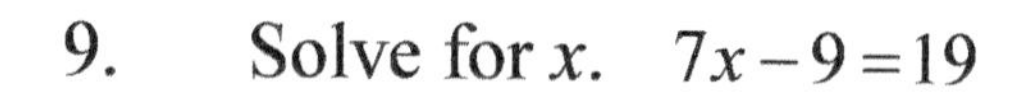

10. $14\frac{2}{5}+12\frac{1}{8}=?$

11. Find the area of a triangle with a base of 16 cm and a height of 3 cm.

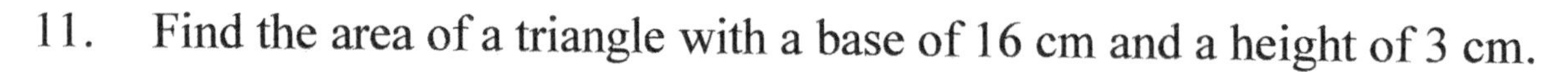

12. $5\leq 11+3x$ Solve and graph the solution.

13. Evaluate x^y+y if $x=3$ and $y=3$.

14. Put these integers in increasing order. –39, 0, –14, –27, 50

15. $0.7-0.1275=?$

16. If the price per pound of lobster goes from \$12 to \$9, what is the percent of change?

17. Find the LCM of $8a^3b^2c$ and $12a^5bc^3$.

18. Solve for x. $\frac{x}{7}=15$

19. Find the values for y in the equation $y=5x+2$ when $x=\{4, 0, -3\}$.

20. Write an algebraic phrase for *nine times a number increased by eight.*

1.	2.	3.	4.
5.	6.	7.	8.
9.	10.	11.	12.
13.	14.	15.	16.
17.	18.	19.	20.

Lesson #5

1. A heptagon has ___________ sides.

2. Find the measurement of the missing angle.

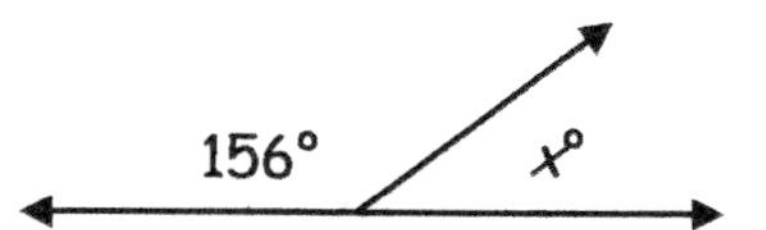

3. Find the area of a parallelogram with a base of 17 mm and a height of 7 mm.

4. Solve for b. $5+b>3$

5. $\frac{3}{8}\times\frac{12}{15}=?$

6. Write the equation of the line that passes through points (3, 5) and (5, 3).

7. $32-5[4+3(2+1)-6]=?$

8. When four friends ate dinner in a restaurant, their bill including tax and tip came to \$25.67. One of them had a \$5.00 discount which was applied to the bill. How much should each person pay? (Round your answer.)

9. $-126-(-79)=?$

10. Write the slope-intercept form of a linear equation.

11. What number is 70% of 25?

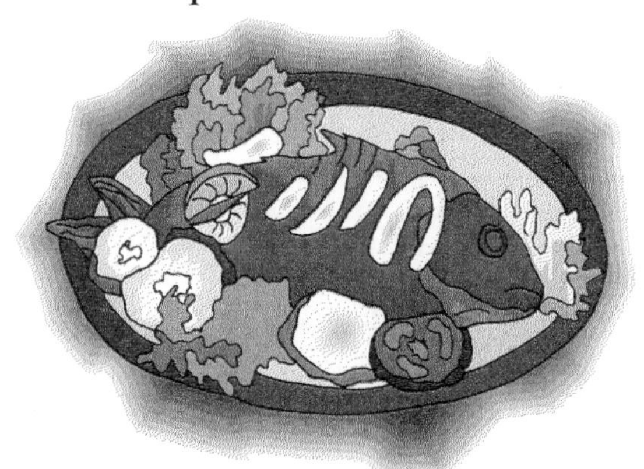

12. Simplify. $\frac{12x^2yz^3}{24x^3y^2z}$

13. Solve for x. $8x-7=17$

14. $-|-19|$

15. Solve for w. $w+44=-78$

16. Solve for x. $13x-26=7x+22$

17. Solve for x. $\frac{9}{7}=\frac{63}{x}$

18. Write 0.55 as a percent and as a reduced fraction.

19. How many centuries are 500 years?

20. Write the coordinates of points A and D.

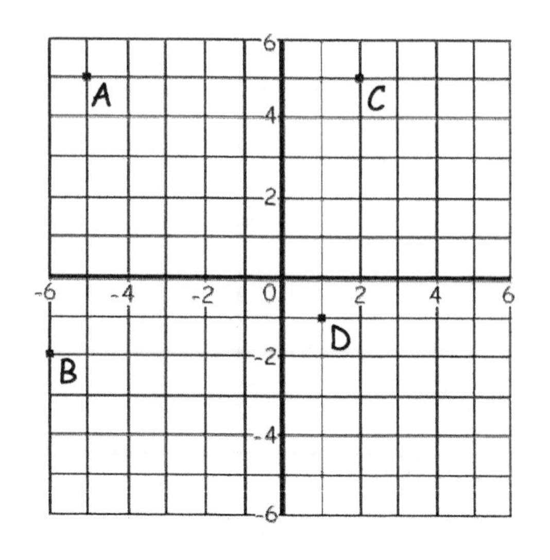

1.	2.	3.	4.
5.	6.	7.	8.
9.	10.	11.	12.
13.	14.	15.	16.
17.	18.	19.	20.

Lesson #6

1. Solve for x. $2x+3x-4=11$

2. $13-5\frac{3}{7}=?$

3. Put these decimals in decreasing order. 6.007, 6.7, 6.07, 6.72

4. $16{,}000-9{,}781=?$

5. Evaluate $x+xy$ if $x=7$ and $y=3$.

6. How many grams are 10 kilograms?

7. Find the area of a circle if the radius is 6 inches.

8. Solve for b. $7b=84$

9. $92+(-37)=?$

10. $50-3[2(6-2)]=?$

11. $3.47+6.9=?$

12. Write $\frac{3}{5}$ as a decimal and as a percent.

13. 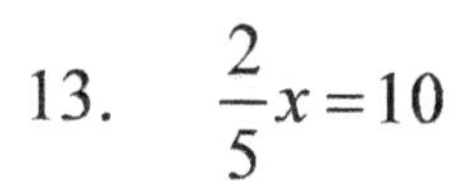 $\frac{2}{5}x=10$

14. Solve for a. $a+33=-90$

15. Find the volume of the cylinder. (See Help Pages for the formula.)

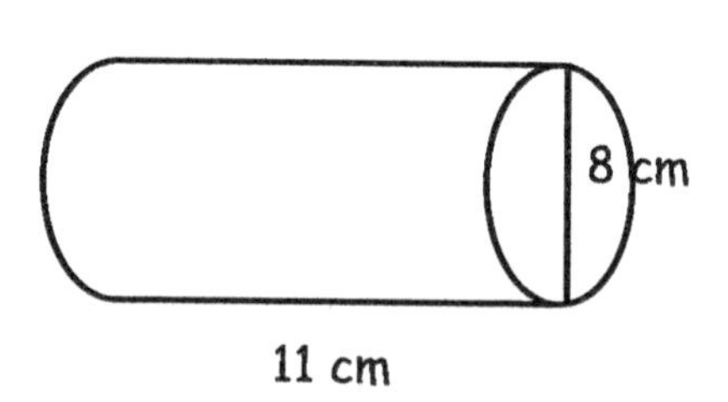

16. Solve for x. $2x+7=7x-18$

17. 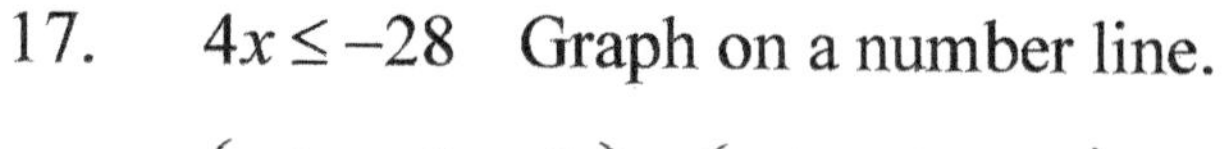 $4x\leq -28$ Graph on a number line.

18. $\begin{pmatrix}-6 & -1 & 7\\ 3 & -2 & -5\end{pmatrix}-\begin{pmatrix}-8 & 6 & -2\\ 0 & -5 & 1\end{pmatrix}=?$

19. 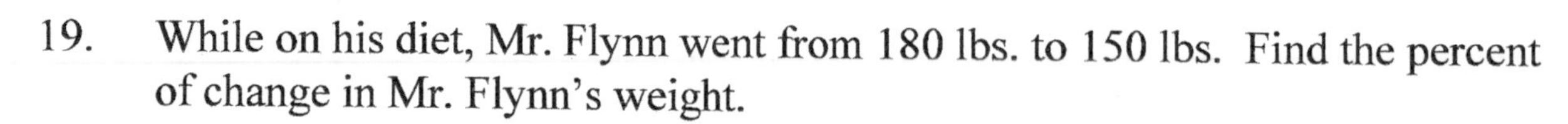 While on his diet, Mr. Flynn went from 180 lbs. to 150 lbs. Find the percent of change in Mr. Flynn's weight.

20. True or false: A line with a positive slope goes upward from left to right.

1.	2.	3.	4.
5.	6.	7.	8.
9.	10.	11.	12.
13.	14.	15.	16.
17.	18.	19.	20.

Lesson #7

1. $0.8 - 0.266 = ?$

2. Which is greater, $\frac{2}{5}$ or 48%?

3. Solve for a. $a - 12 = 52$

4. Write an algebraic phrase to represent *five less than the quotient of a number and six.*

5. Find the GCF of $10ab^2c^2$ and $15a^2bc^2$.

6. Solve for x. $4x + 16 = 28$

7. Find the area of the trapezoid.

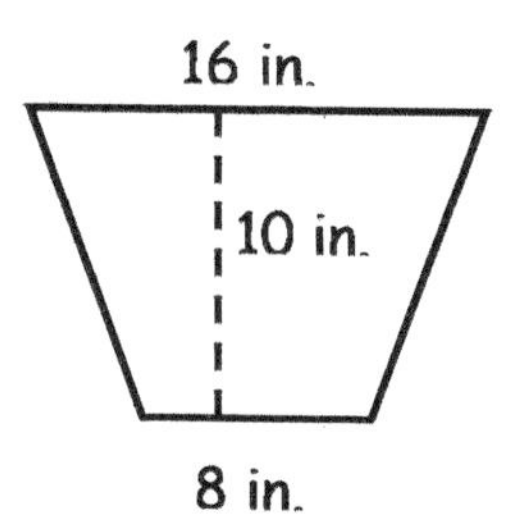

8. $-93 + (-54) = ?$

9. $2x + 4 \geq 2$ Graph the solution on a number line.

10. $-4(-2)(9) = ?$

11. 60 is what percent of 80?

12. $23\frac{1}{4} + 18\frac{2}{5} = ?$

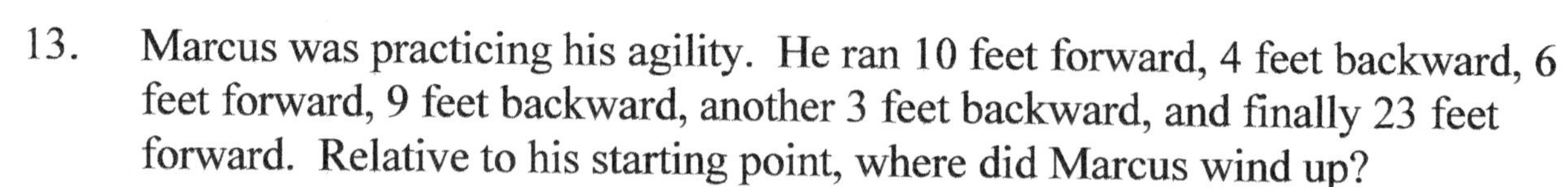

13. Marcus was practicing his agility. He ran 10 feet forward, 4 feet backward, 6 feet forward, 9 feet backward, another 3 feet backward, and finally 23 feet forward. Relative to his starting point, where did Marcus wind up?

14. Write 68,270,000 in scientific notation.

15. How many pounds are 7 tons?

16. $1.402 \div 2 = ?$

17. $\frac{3x - 2}{5} = -7$ (Hint: Multiply both sides by 5.)

18. Simplify. $4(3x - 2y + 5) + 2(4x - 7)$

19. Evaluate $\frac{x}{y} + xy$ if $x = 8$ and $y = 2$.

20. On her social studies test, Rachael got 15 wrong out of 25 questions. What percent of the questions did she miss?

1.	2.	3.	4.
5.	6.	7.	8.
9.	10.	11.	12.
13.	14.	15.	16.
17.	18.	19.	20.

Lesson #8

1. Write 0.000478 in scientific notation.

2. $0.008 \times 0.03 = ?$

3. Solve for x. $\frac{11x-3}{6} = 5$

4. What is the P(3, 1, 5) on 3 rolls of a die?

5. Solve for x. $6x + 12 = 24$

6. Solve for x. $\frac{4}{5}x = 20$

7. Evaluate $x(y+6)$ when $x = 2$ and $y = 3$.

8. 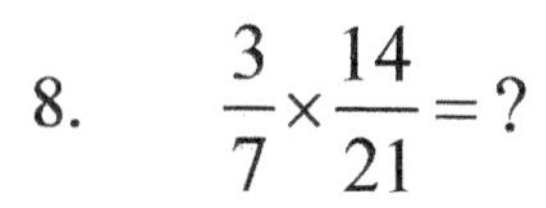 $\frac{3}{7} \times \frac{14}{21} = ?$

9. Find the volume of a rectangular prism if its length is 12 cm, its width is 6 cm, and its height is 3 cm.

10. Solve for x. $\frac{x}{9} = 10$

11. Find the values for y in the equation $y = 4x$ when $x = \{1, 0, -3\}$.

12. $\begin{pmatrix} 4 & 0 \\ -3 & -5 \end{pmatrix} + \begin{pmatrix} 5 & -6 \\ -2 & 3 \end{pmatrix} = ?$

13. Solve for a. $6a + 10 = 22a - 22$

14. Solve for x. $5 - 2x \leq 3 - x$

15. $5 + 3[2(5-2) + 4] = ?$

16. Find the slope of a line passing through points (9, –2) and (3, 4).

17. Write an algebraic phrase to represent *the sum of a number and twelve.*

18. Mrs. Grady pays $258 each month for medical insurance. Next month, the bill will reflect a rate increase of 4%. How much will she have to pay for medical insurance next month?

19. A triangle with no congruent sides is a(n) ____________ triangle.

20. Write $\frac{5}{8}$ as a decimal.

1.	2.	3.	4.
5.	6.	7.	8.
9.	10.	11.	12.
13.	14.	15.	16.
17.	18.	19.	20.

Lesson #9

1. The O'Connor Family's old boat was 16 ft. long. Their new boat is 24 ft. long. What is the percent of change?

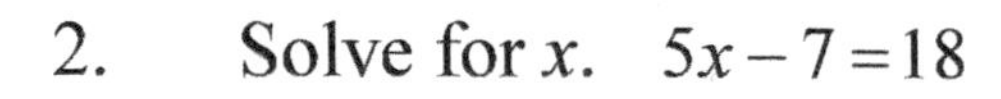

2. Solve for x. $5x - 7 = 18$

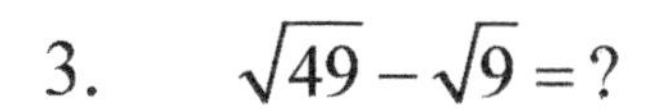

3. $\sqrt{49} - \sqrt{9} = ?$

4. A decagon has __________ sides.

5. The slope of a vertical line is __________.

6. $19\frac{4}{7} - 3\frac{1}{7} = ?$

7. What number is 60% of 80?

8. $8{,}000{,}000 - 2{,}477{,}653 = ?$

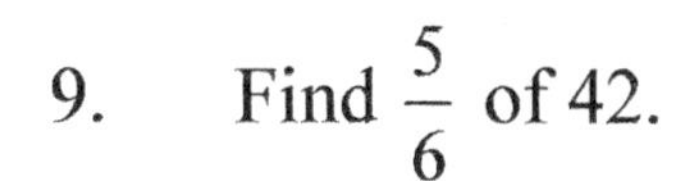

9. Find $\frac{5}{6}$ of 42.

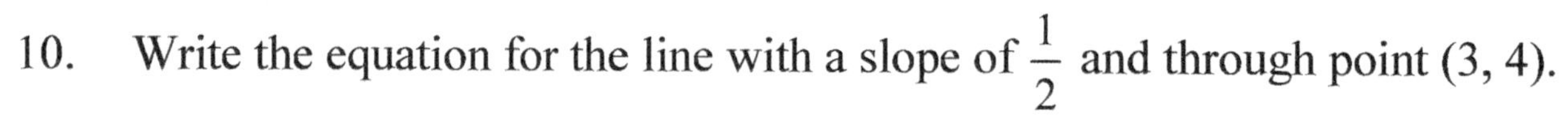

10. Write the equation for the line with a slope of $\frac{1}{2}$ and through point (3, 4).

11. Solve for x. $\frac{21}{12} = \frac{7}{x}$

12. Solve for x. $\frac{x}{6} - 5 = 15$

13. $-7(-5) = ?$

14. Simplify. $3(4a - 5b + 6) - 2a$

15. Evaluate $a(b + 5)$ when $a = 4$ and $b = 3$.

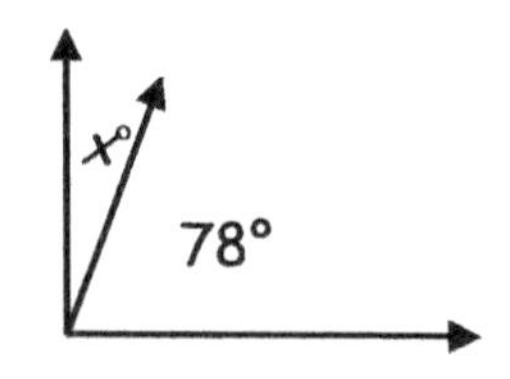

16. Find the missing measurement, x.

17. Mr. Thompson ordered 425 pens with his business name imprinted on them and 500 pens with no imprint. What percent of the total number of pens were imprinted? Round your answer to the nearest percent.

18. $-114 - (-67) = ?$

19. $24 \div 4 + 6 \cdot 2 - 3 = ?$

20. Round 6.274 to the nearest tenth.

1.	2.	3.	4.
5.	6.	7.	8.
9.	10.	11.	12.
13.	14.	15.	16.
17.	18.	19.	20.

Lesson #10

1. $6\frac{2}{3}+4\frac{1}{9}=?$

2. Write the formula for finding the surface area of a rectangular prism.

3. $\begin{pmatrix}7 & -2\\ 8 & 0\end{pmatrix}-\begin{pmatrix}5 & -5\\ -3 & 7\end{pmatrix}=?$

4. Write the equation of the line that passes through points (25, 100) and (15, 120).

5. Simplify. $\frac{16a^4b^2c^3}{24a^2b^2c^2}$

6. Solve for x. $3x+12>21-2x$

7. Solve for x. $5x+3=18$

8. How many ounces are in 6 pounds?

9. Find x. $\frac{1}{9}x-7=20$

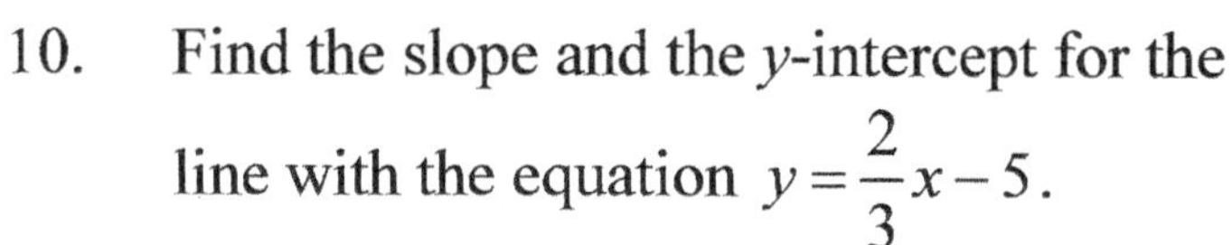

10. Find the slope and the y-intercept for the line with the equation $y=\frac{2}{3}x-5$.

11. Write 1.25×10^4 in standard form.

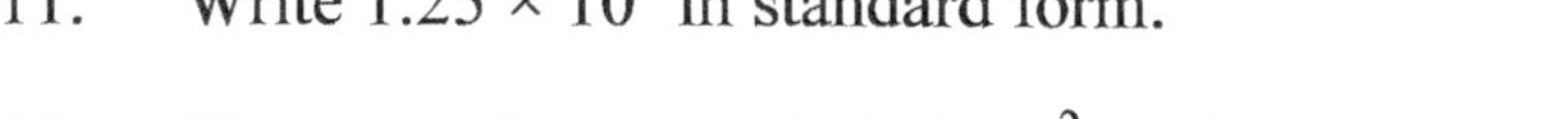

12. The area of a square is 100 cm^2. What is the length of each side?

13. 375,218 + 998,667 = ?

14. Find the average of 65, 75 and 25.

15. $(-7)^3=?$

16. The slope of a horizontal line is ________.

17. Solve for x. $9x=-108$

18. $24.36 \div 0.04=?$

19. $99+(-32)=?$

20. Give the coordinates for points A, B and C on the graph.

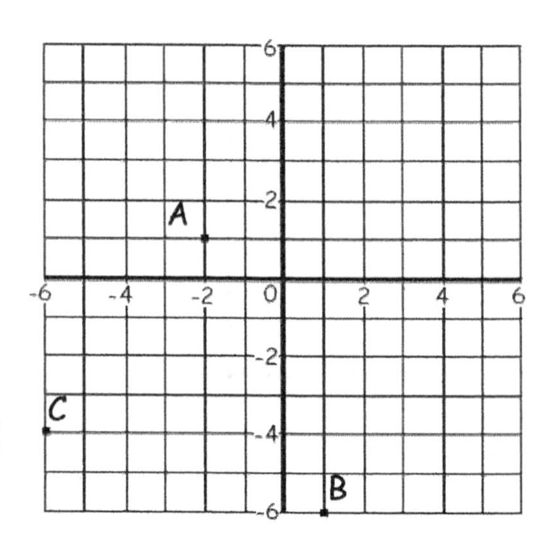

1.	2.	3.	4.
5.	6.	7.	8.
9.	10.	11.	12.
13.	14.	15.	16.
17.	18.	19.	20.

Lesson #11

1. Solve for a. $4a-32=-12$

2. Graph the solution to $g-3>2$ on a number line.

3. What are the slope and the y-intercept for a line with the equation $y=3x-7$?

4. Solve for x. $2x-2>4$

5. Solve for a. $\frac{1}{6}a+7=12$

6. How many feet are in 5 yards?

7. Solve for x. $\frac{2x}{5}=18$

8. $2.5\times1.6=?$

9. Find the circumference of a circle with a diameter of 12 yds.

10. $\frac{4}{5}\times\frac{15}{24}=?$

11. If $x=3$ and $y=2$, what is the value of $4x+y$?

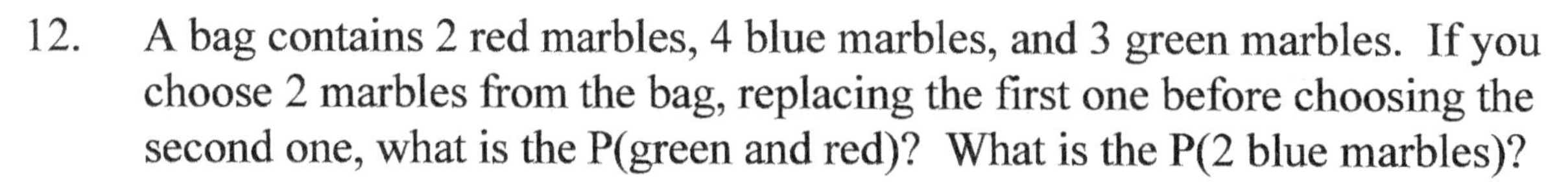

12. A bag contains 2 red marbles, 4 blue marbles, and 3 green marbles. If you choose 2 marbles from the bag, replacing the first one before choosing the second one, what is the P(green and red)? What is the P(2 blue marbles)?

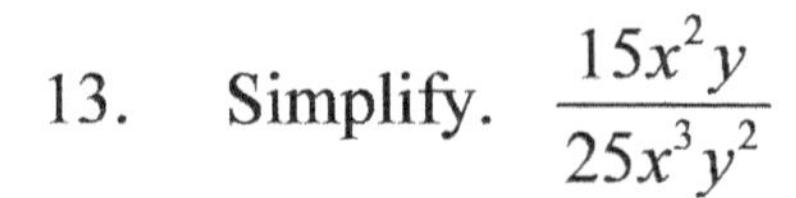

13. Simplify. $\frac{15x^2y}{25x^3y^2}$

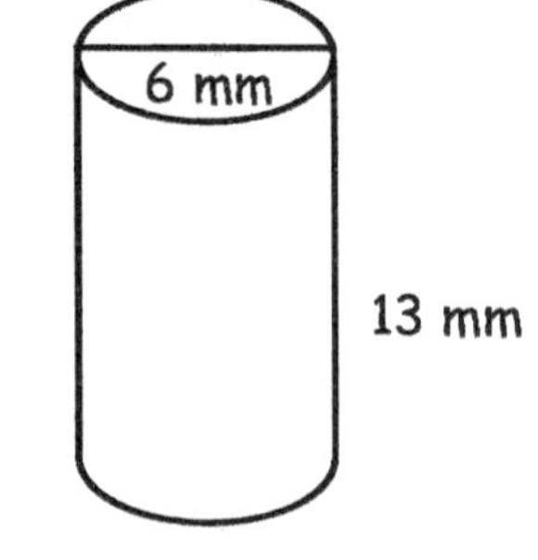

14. Find the volume of the cylinder.

15. Solve for x. $\frac{x}{8}-9=20$

16. $(-5)^4=?$

17. Write 0.32 as a percent and as a reduced fraction.

18. Solve for x. $9x+3=2x-18$

19. Solve for x. $\frac{5}{6}x-\frac{4}{6}x+2=11$

20. What percent of 90 is 27?

1.	2.	3.	4.
5.	6.	7.	8.
9.	10.	11.	12.
13.	14.	15.	16.
17.	18.	19.	20.

Lesson #12

1. $15-2[4+2(5)]=?$

2. A triangle with 2 congruent sides is a(n) __________ triangle.

3. What value of x makes this equation true? $9x+63=-81$

4. $-33-(-71)=?$

5. Solve for x. $4x-10=6x+12$

6. Write an algebraic expression to represent *six more than twice a number.*

7. Solve for a. $8a-2a+12=42$

8. Find the LCM of $9a^3b^2$ and $15ab^4$.

9. Solve for c. $c+196=243$

10. Determine the area of the triangle.

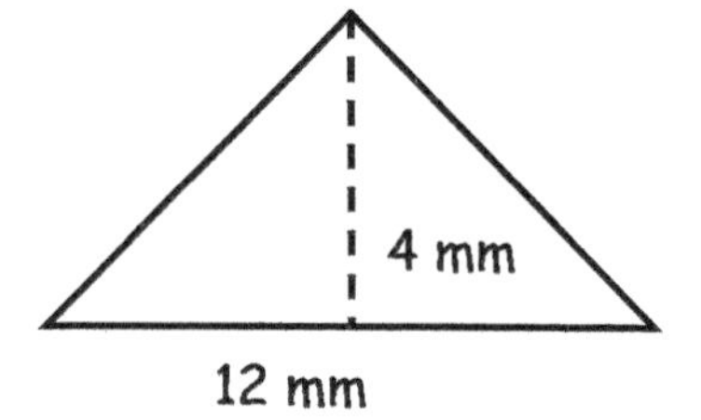

11. Mark saw a tent which was \$16 more than twice the cost of a sleeping bag. The tent was \$148. What was the cost of the sleeping bag? Write an algebraic equation and solve for x.

12. Solve the equation, $y=5x$, for y when $x=\{-4, 0, 2\}$.

13. $4\frac{3}{8}+2\frac{1}{2}=?$

14. Evaluate $\frac{6a}{b}-5$ when $a=3$ and $b=2$.

15. Write $\frac{1}{8}$ as a decimal.

16. How many cups are in 8 pints?

17. $4,076-1,964=?$

18. $-5(11)=?$

19. Round 87,354,277 to the nearest million.

20. Solve the inequality and graph on a number line. $\frac{x}{7}>5$

1.	2.	3.	4.
5.	6.	7.	8.
9.	10.	11.	12.
13.	14.	15.	16.
17.	18.	19.	20.

Lesson #13

1. Solve for a. $12a+8=6a$

2. $0.009\times 0.004=?$

3. What value(s) of x make this a true statement? $2<-8x$

4. Solve for a. $12a+8=6a-10$

5. How many feet are in 3 miles?

6. $62\frac{1}{9}-37\frac{7}{9}=?$

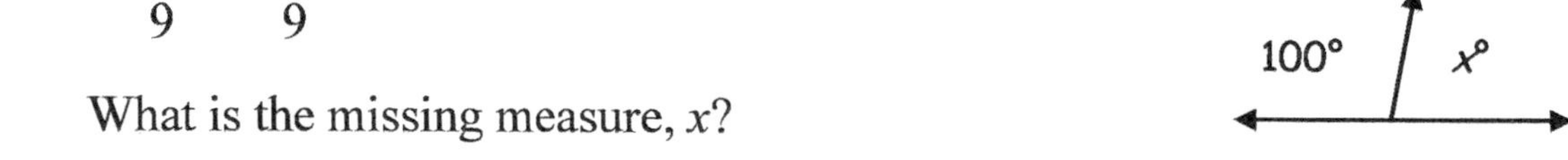

7. What is the missing measure, x?

8. $-4<x-5\leq -1$ Find the solution and graph it on a number line.

9. Write 45% as a decimal and as a reduced fraction.

10. Solve for b. $\frac{3}{5}=\frac{b}{75}$

11. $35+5(4+3)-2\cdot 5=?$

12. The population of Oak Grove dropped from 15,000 to 12,000. By what percent did the population change?

13. Evaluate $ab-c$ when $a=3$, $b=2$, and $c=4$.

14. Solve for a. $a+29=-66$

15. Find the equation of the line that passes through points (3, 4) and (–2, 1).

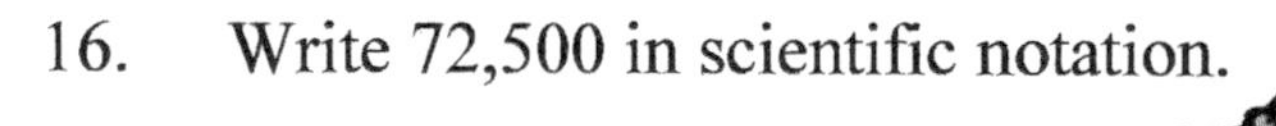

16. Write 72,500 in scientific notation.

17. $72+(-48)=?$

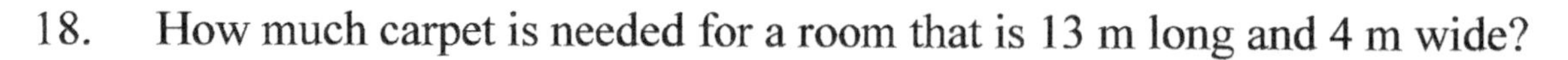

18. How much carpet is needed for a room that is 13 m long and 4 m wide?

19. What is the value of x? $\frac{4}{5}x-\frac{3}{5}x+9=14$

20. $\begin{pmatrix} 5 & 0 & -3 \\ 8 & -1 & -2 \end{pmatrix}-\begin{pmatrix} 8 & -4 & -6 \\ 5 & -2 & 4 \end{pmatrix}=?$

1.	2.	3.	4.
5.	6.	7.	8.
9.	10.	11.	12.
13.	14.	15.	16.
17.	18.	19.	20.

Lesson #14

1. Write the equation of the line with a slope of –3 and through point (5, –8).

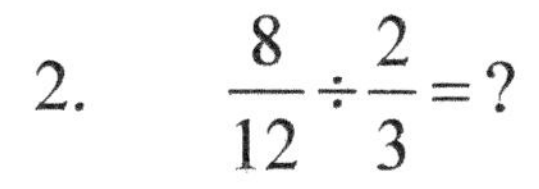

2. $\frac{8}{12} \div \frac{2}{3} = ?$

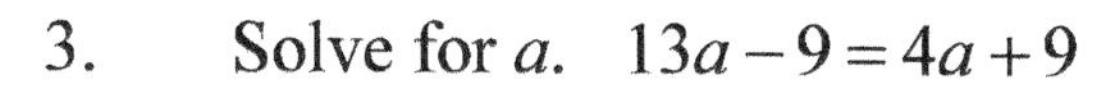

3. Solve for a. $13a - 9 = 4a + 9$

4. How many decades are 80 years?

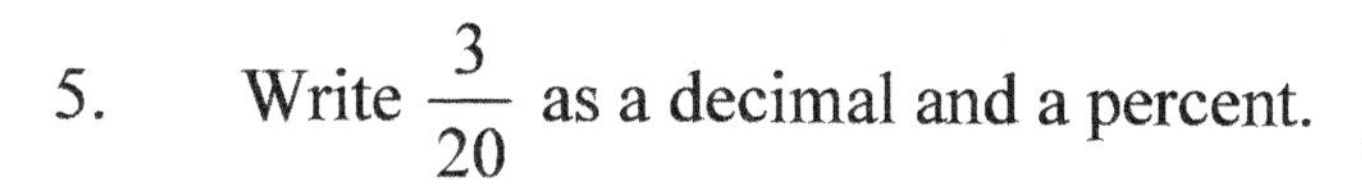

5. Write $\frac{3}{20}$ as a decimal and a percent.

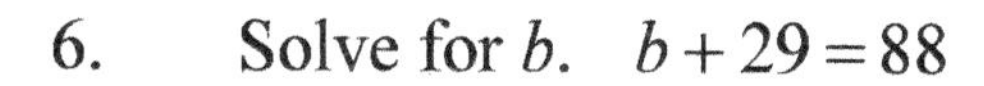

6. Solve for b. $b + 29 = 88$

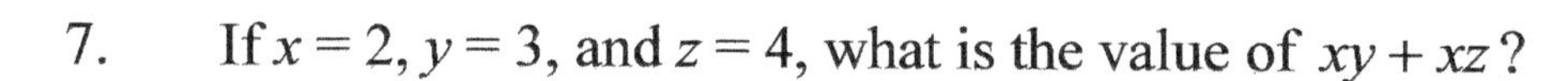

7. If $x = 2$, $y = 3$, and $z = 4$, what is the value of $xy + xz$?

8. Solve the equation, $y = 7x + 2$, for y when $x = \{0, -3, 1\}$.

9. Simplify. $7(3a + 4b + 5) - 6a$

10. Simplify. $\frac{14a^3b^2}{21ab}$

11. Find the volume of a cube if the sides each measure 9 inches.

12. Solve for x. $5x - 7 = 18$

13. The scouts had a bake sale. They cut eight 12" × 6" cakes into slices that each measured $\frac{2}{3}$" × 3". How many slices were there?

14. Solve for x. $\frac{x}{8} + 5 = 14$

15. What is the P(H, H, H, T, T, T) on 6 flips of a coin?

16. In a marsh, the ratio of ducks to birds was 7 to 9. If there were 119 ducks, how many birds were in the marsh?

17. Find the solution and graph it on a number line. $2x + 2 > 4$

18. What number is 25% of 80?

19. Find the median and the mode of 23, 48, 19, 66 and 23.

20. Solve for x. $\frac{2x-1}{5} = 3$ (Hint: Multiply both sides by 5.)

1.	2.	3.	4.
5.	6.	7.	8.
9.	10.	11.	12.
13.	14.	15.	16.
17.	18.	19.	20.

Lesson #15

1. Solve for x. $\frac{7}{8}x = 14$

2. $10\frac{1}{3} + 5\frac{1}{5} = ?$

3. $45{,}013 - 26{,}778 = ?$

4. Write 1.48×10^{-3} in standard form.

5. Write $\frac{2}{50}$ as a decimal and as a percent.

6. $\frac{-60}{-3} = ?$

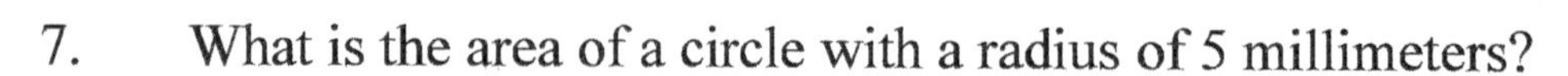

7. What is the area of a circle with a radius of 5 millimeters?

8. Solve for x. $\frac{x}{4} - 12 = 20$

9. What is the value of $5ab + 2b$ when $a = 3$ and $b = 4$?

10. Find $\frac{4}{5}$ of 35.

11. In a clay pot, Monica added $2\frac{1}{3}$ cups of sand to $5\frac{1}{4}$ cups of dirt. How much of the soil mixture did Monica make?

12. $-|-92| = ?$

13. Solve for a. $3a + 8 = 9a - 16$

14. $4[5 + 2(7 + 2) - 5] + 2^2 = ?$

15. Simplify. $(-6)^4 = ?$

16. A line passing through points (–2, 1) and (5, 7) has what slope?

17. Solve for x. $3(x + 2) = 12$

18. How many inches are in 3 yards?

19. Write an algebraic sentence for *ten times a number divided by four.*

20. True or false. A line with negative slope goes downward from right to left.

1.	2.	3.	4.
5.	6.	7.	8.
9.	10.	11.	12.
13.	14.	15.	16.
17.	18.	19.	20.

Lesson #16

1. What is the slope of the line passing through points (– 4, –5) and (–9, 1)?

2. $\frac{5}{6} \times \frac{12}{20} = ?$

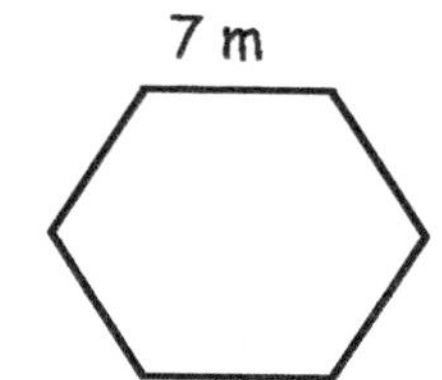

3. What is the perimeter of the hexagon?

4. $-44 + (-86) = ?$

5. Of the 50 students in the sixth grade, 16 are girls. What percent are girls?

6. Solve for b. $b + 24 = -77$

7. $0.72 - 0.5498 = ?$

8. What percent of 80 is 64?

9. A triangle with all sides congruent is a(n) __________ triangle.

10. Solve for m. $8 \leq 4 - m$

11. The cost of a pair of jeans went from $15 to $24. By what percent did the cost increase?

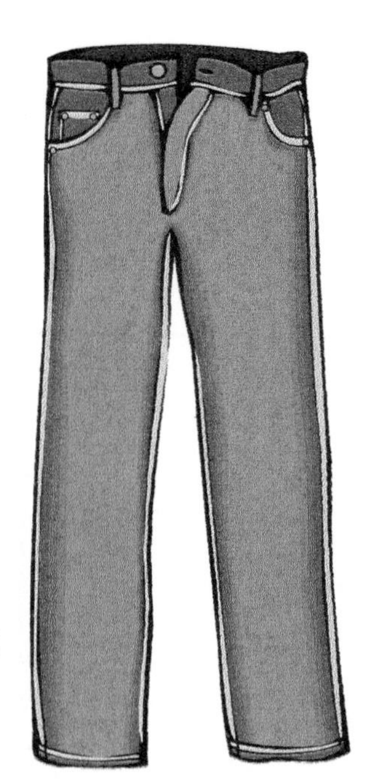

12. $55.8 \div 2 = ?$

13. Solve for x. $\frac{x}{9} = -3$

14. $\frac{8}{12} \div \frac{4}{12} = ?$

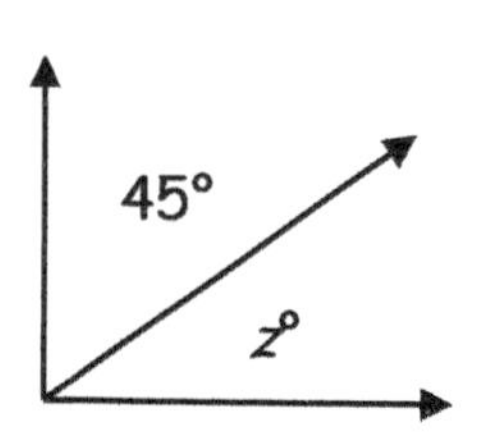

15. What is the missing measure, z?

16. What value of x makes the equation true? $2x + 4 = x - 6$

17. Write 0.0000064 in scientific notation.

18. Solve for x. $7x + 3 = -18$

19. Melissa practiced $12\frac{1}{2}$ measures of a musical piece. The entire piece was 21 measures long. How many measures did she have left to practice?

20. Evaluate $b(8 - a) + c$ when $a = 2$, $b = 4$, and $c = 3$.

1.	2.	3.	4.
5.	6.	7.	8.
9.	10.	11.	12.
13.	14.	15.	16.
17.	18.	19.	20.

Lesson #17

1. $6 \div 1.5 = ?$

2. Write an equation of the line through point (1, 2) with a slope of –3.

3. $18 - 12\frac{5}{6} = ?$

4. How many quarts are in 5 gallons?

5. Solve the system $y = 2x - 3$ and $y = x - 1$.

6. Solve for c. $c - 62 = 97$

7. The slope of a horizontal line is ______________.

8. Solve for x. $\frac{x}{14} + 3 = 11$

9. What value of x makes the equation true? $5x + 6 = 2x + 18$

10. What is the circumference of a circle with a diameter of 14 cm?

11. $-325 + (-136) = ?$

12. Write 3.445×10^5 in standard form.

13. Determine the solution set for $3a + 5 < -4$.

14. $8\frac{1}{3} \times 3\frac{3}{5} = ?$

15. $\sqrt{144} + 4^3 = ?$

16. Write $\frac{3}{8}$ as a decimal.

17. If $a = 2$ and $b = 3$, what is the value of $ab - a - b$?

18. Solve for x. $\frac{36}{x} = \frac{16}{24}$

19. $\begin{pmatrix} 6 & -4 \\ 5 & 0 \end{pmatrix} - \begin{pmatrix} 2 & -7 \\ -5 & 6 \end{pmatrix} = ?$

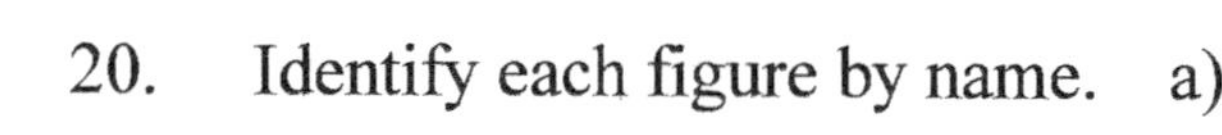

20. Identify each figure by name. a) 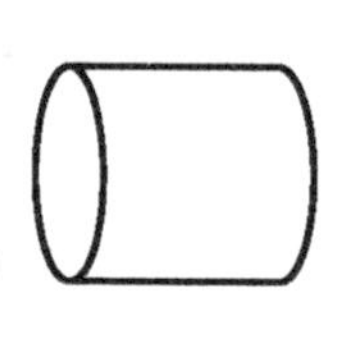 b) 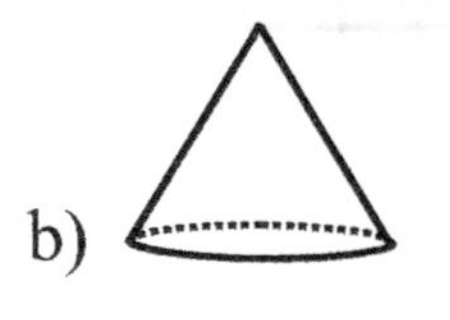 c) 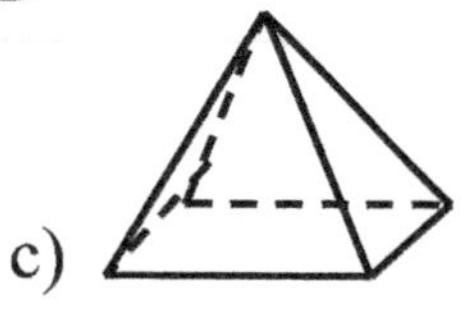

1.	2.	3.	4.
5.	6.	7.	8.
9.	10.	11.	12.
13.	14.	15.	16.
17.	18.	19.	20.

Lesson #18

1. Solve the system of equations. $y = -x + 4$ and $y = 2x + 1$

2. $\dfrac{32}{-4} = ?$

3. Solve for *n*. $\dfrac{2}{3}n = 12$

4. What value of *x* makes the equation true? $x - 24 = 98$

5. 20% of 30 is what number?

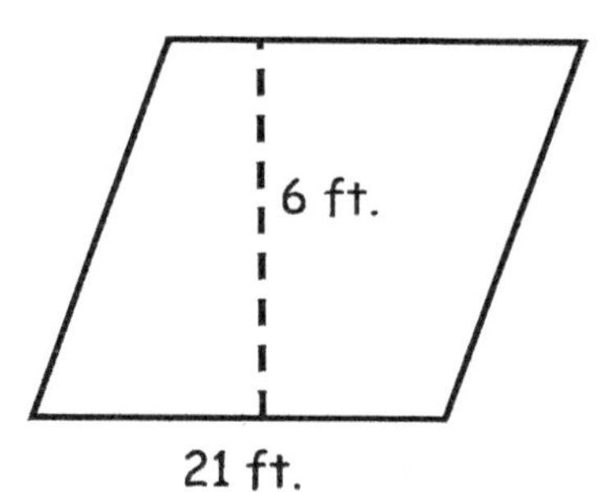

6. $\sqrt{81} + 3\left[2(8-3)+1\right] = ?$

7. Calculate the area of the parallelogram.

8. Solve for *a*. $10a - 4 = -94$

9. 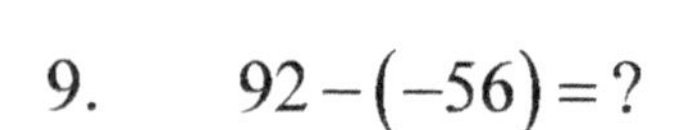
$92 - (-56) = ?$

10. You have three \$1 bills, two \$5 bills, and a \$20 bill in your pocket. You choose two bills without replacing them. What is the P(\$1 and \$20)?

11. What is the value of *a* if $12a - 7 = 4a + 9$?

12. Find the values for *y* in the equation $y = 7x - 8$ when $x = \{0, -1, 2\}$. Graph the solution on a coordinate graph.

13. What algebraic phrase means *the sum of a number and seven less three*?

14. $4x + 3 < -5$ or $-2x + 7 < 1$ Graph the solution on a number line.

15. $-7^3 = ?$

16. Write 63,400,000 in scientific notation.

17. $\dfrac{8}{10} \div \dfrac{4}{5} = ?$

18. $3.5 \times 0.7 = ?$

19. Solve for *a*. $\dfrac{1}{8}a - 11 = -21$

20. A line passes through the points (4, –1) and (4, 7). What is its slope?

1.	2.	3.	4.
5.	6.	7.	8.
9.	10.	11.	12.
13.	14.	15.	16.
17.	18.	19.	20.

Lesson #19

1. A \$10 T-shirt is on sale for \$4. What is the percent of the discount?

2. Write the slope and the y-intercept for the line if $y = 3x - 5$.

3. On a car lot, 32 of 80 cars were front wheel drive. What percent of the cars were not front wheel drive?

4. The measure of angle x is ________.

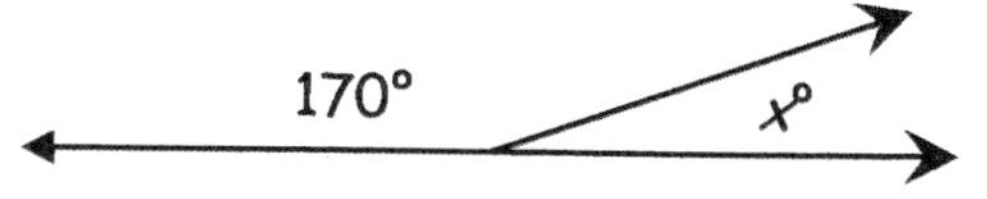

5. $-3(-4)(-2) = ?$

6. Solve the system using substitution. $y = 2x + 1$, $y = x + 3$

7. Solve for x. $\frac{x}{7} + 15 = 25$

8. Evaluate $m + n - mn$ if $m = 5$ and $n = 2$.

9. Determine the area of the trapezoid.

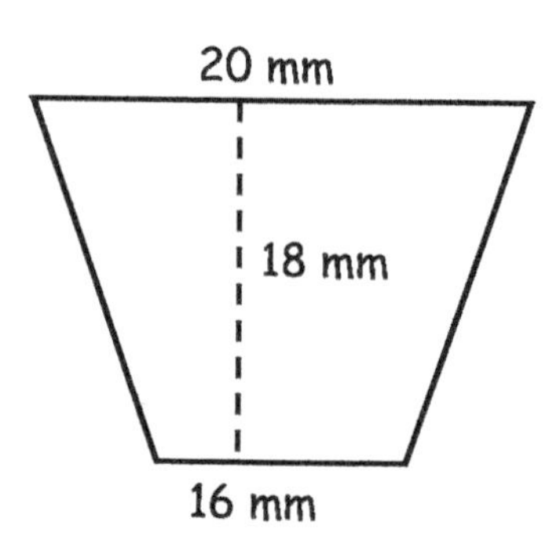

10. A and B are independent events. If the $P(A) = \frac{1}{3}$ and the $P(B) = \frac{3}{5}$, what is the P(A and B)?

11. How many pints are in 5 quarts?

12. Graph the solution to the inequality on a number line. $b - 3 \geq 4$

13. $19 - (-88) = ?$

14. Solve for a. $20a + 14 = 50 + 8a$

15. Write the formula for finding the volume of a cylinder.

16. $1.6 - 0.887 = ?$

17. Solve for c. $c + 27 = -76$

18. A nonagon has ________ sides.

19. $23\frac{1}{7} - 16\frac{5}{7} = ?$

20. Write the slope-intercept form of a linear equation.

1.	2.	3.	4.
5.	6.	7.	8.
9.	10.	11.	12.
13.	14.	15.	16.
17.	18.	19.	20.

Lesson #20

1. $-42+(-28)+(-12)=?$

2. Evaluate $3a+2b-ab$ when $a=4$ and $b=3$.

3. Simplify. $\dfrac{8a^3b^2c}{12ab^3c^4}$

4. Use substitution to solve the system of equations: $\begin{aligned} x+y&=6 \\ x&=-3y \end{aligned}$

5. $\begin{pmatrix} 8 & -7 \\ 3 & -5 \end{pmatrix}+\begin{pmatrix} 4 & 0 \\ -2 & -4 \end{pmatrix}=?$

6. Write 2.085×10^5 in standard form.

7. Solve for x. $4x-2=10$

8. $[8+(2\cdot 4)]\cdot 3=?$

9. Find the range of 86, 31, 16, 25 and 90.

10. How many degrees are in a circle?
In a straight line?

11. Solve for x. $2(x-3)=8$

12. What is the value of x in $\left(\dfrac{4}{9}x=36\right)$?

13. What are the slope and the y-intercept of the line if $y=\dfrac{3}{5}x+6$?

14. What value of x makes the equation true? $5x-3=2x+12$

15. Write the equation of the line with a slope of $\dfrac{-3}{4}$, passing through point (0, 2).

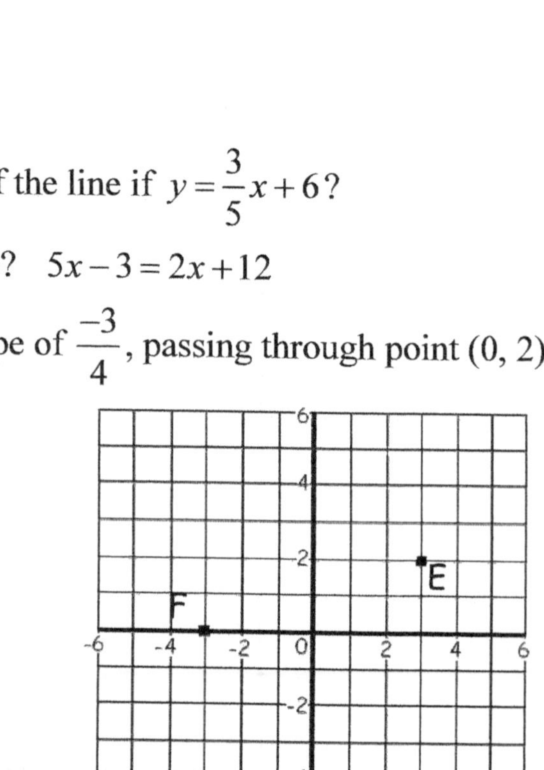

16. Solve for x. $\dfrac{x+4}{3}=-1$

17. Solve for a. $a+14=41$

18. $1\dfrac{1}{2}\cdot\dfrac{2}{3}=?$

19. Solve for x. $\dfrac{x}{3}=11$

20. Give the coordinates for points E and F.

1.	2.	3.	4.
5.	6.	7.	8.
9.	10.	11.	12.
13.	14.	15.	16.
17.	18.	19.	20.

Lesson #21

1. $-142-(-79)=?$

2. Write the equation for a line passing through point (0, 3) with a slope of 1.

3. Write 0.000089 in scientific notation.

4. $2.75+9.964=?$

5. Calculate the area of a circle if its radius is 4 mm.

6. How many ounces are in 4 pounds?

7. Use substitution to solve the system. $x=-2$
 $3x-2y=4$

8. Solve for x. $\frac{4}{21}=\frac{x}{84}$

9. What is the value of x in $2x-7=15$?

10. $\begin{pmatrix} 5 & -3 & 2 \\ 9 & -7 & 0 \end{pmatrix}+\begin{pmatrix} 6 & -7 & -4 \\ -3 & -5 & 2 \end{pmatrix}=?$

11. Graph the solution set on a number line. $12x \le 3x+27$

12. Last year, Marti's locker was 12 inches wide. This year her locker is 19 inches wide. By what percent has the width increased? (Round to the nearest whole number.)

13. Solve for a. $a+25=70$

14. What value of a makes the equation true? $6a-14=13+3a$

15. $-8(9)=?$

16. If you roll a die twice, what is the probability of rolling a 1 followed by a 6?

17. Solve for x. $\frac{x}{12}+9=+15$

18. What is the slope of a horizontal line?

19. What number is 70% of 60?

20. Find the solution set. $t-5 \le -3$

1.	2.	3.	4.
5.	6.	7.	8.
9.	10.	11.	12.
13.	14.	15.	16.
17.	18.	19.	20.

Lesson #22

1. Write the formula for finding the area of a triangle.

2. $246-(-78)=?$

3. $2b+3<7$ Graph on a number line.

4. $\dfrac{64}{-4}=?$

5. How many grams are in 7 kilograms?

6. A heptagon has _________ sides.

7. What is the measure of the angle that is supplementary to the angle shown?

8. Solve for n. $8n-3=13$

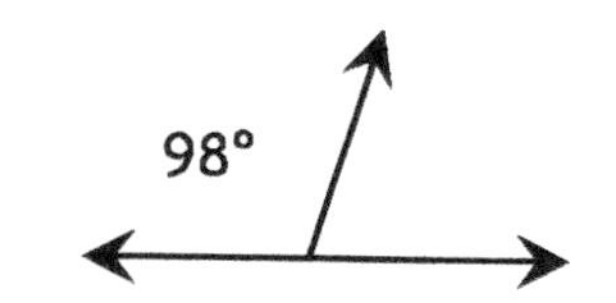

9. Solve for x. $\frac{1}{7}x+5=14$

10. $25+3[2+3\cdot4-1]=?$

11. Larry has 4 bookcases, each $2\frac{1}{3}$ feet wide. How many feet of wall space will all 4 bookcases take up if they are placed end to end?

12. Evaluate $\dfrac{ab}{c}+a$ if $a=2$, $b=9$, and $c=3$.

13. Solve the system by substitution. $x-y=20$, $2x+3y=0$

14. Write the equation for the line through point (4, 6), with a slope of –5.

15. Solve for b. $b+26=-64$

16. Which is greater, $\dfrac{9}{50}$ or 28%?

17. Simplify. $7(2a+4b-3)+2(3a-5)$

18. $0.0006\times0.03=?$

19. The line passing through points (2, 5) and (4, 8) has what slope?

20. Round 82.6345 to the nearest thousandth.

1.	2.	3.	4.
5.	6.	7.	8.
9.	10.	11.	12.
13.	14.	15.	16.
17.	18.	19.	20.

Lesson #23

1. Solve the system using elimination. $$\begin{aligned}5x-6y&=-32\\3x+6y&=48\end{aligned}$$

2. Caryn bought a new necklace that was priced at $375. The sales tax rate was 7%. What was the total cost of the necklace?

3. $500{,}000-235{,}786=?$

4. Write the slope and the y-intercept for $y=-2x-9$.

5. $-34+(-18)+(-26)=?$

6. $7\frac{3}{4}+6\frac{1}{2}=?$

7. Write $\frac{3}{20}$ as a decimal and as a percent.

8. Put these decimals in decreasing order.

   3.25 3.5 3.205 3.025

9. Solve for b. $b-136=255$

10. Simplify. $\dfrac{18a^2bc^3}{24abc^3}$

11. What is the value of $2x+y-4$ when $x=3$ and $y=2$?

12. Calculate the surface area of the prism.

6 m
4 m
14 m

13. Solve for x. $3x-5=10$

14. $\frac{5}{9}\times\frac{12}{25}=?$

15. $9\cdot4+36\div3-2^2=?$

16. The line passing through points (1, 6) and (7, 3) has what slope?

17. Write 9.34×10^{-5} in standard form.

18. What are the values for y in the equation $y=-6x$ when $x=\{0, -27, 7\}$?

19. Graph the inequality on a number line. $x-5\geq-13$

20. Solve for p. $|p-9|=3$

1.	2.	3.	4.
5.	6.	7.	8.
9.	10.	11.	12.
13.	14.	15.	16.
17.	18.	19.	20.

Lesson #24

1. Draw perpendicular lines.

2. How many degrees are in a circle?

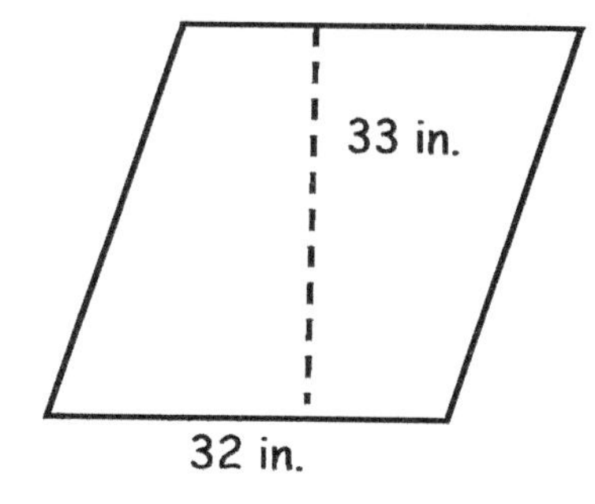

3. 80% of what number is 20?

4. $-425 + 276 = ?$

5. Find the area of the parallelogram.

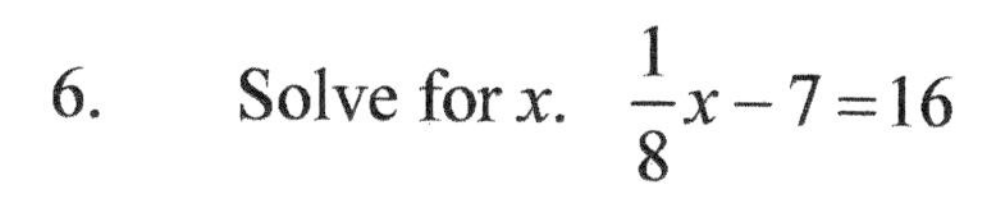

6. Solve for x. $\frac{1}{8}x - 7 = 16$

7. A ten-sided shape is called a(n) ________________.

8. Find the GCF of $10x^2y^3z^4$ and $12xy^2z^3$.

9. Write an algebraic phrase for *six times a number plus three times another number.*

10. The Celsius boiling point of water is ______________.

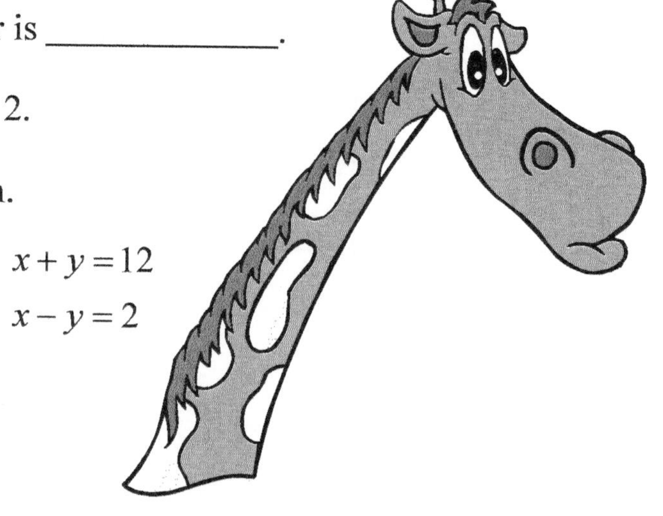

11. Evaluate $x^y + xy$ if $x = 5$ and $y = 2$.

12. Write 8.934×10^6 in standard form.

13. Solve the system by elimination. $x + y = 12$, $x - y = 2$

14. $20 + 2 \cdot 5 + 3[(8-5)-1] = ?$

15. Solve for x. $5x - 6 = 24$

16. $\begin{pmatrix} 2 & -7 \\ 5 & 0 \end{pmatrix} - \begin{pmatrix} 3 & -4 \\ -1 & 6 \end{pmatrix} = ?$

17. $2x + 3 < 7$ Graph the inequality on a number line.

18. $(-3)^5 = ?$

19. Solve for b. $b - 22 = 40$

20. Give the points with coordinates (–3, 0) and (3, 2).

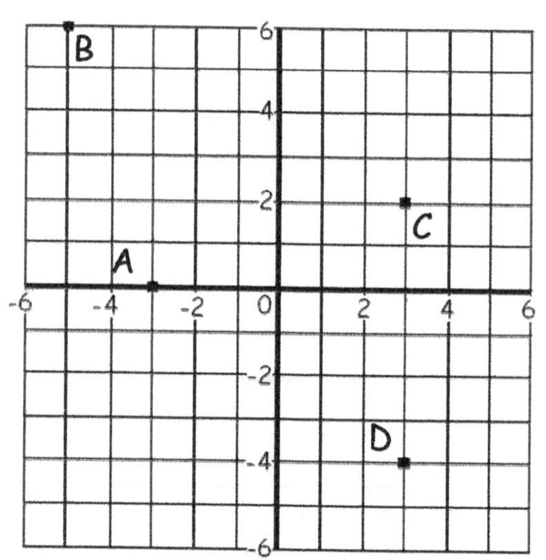

1.	2.	3.	4.
5.	6.	7.	8.
9.	10.	11.	12.
13.	14.	15.	16.
17.	18.	19.	20.

Lesson #25

1. $\frac{8}{9} \times \frac{18}{24} = ?$

2. Write the equation of the line passing through points (7, 3) and (2, 2).

3. $-16+12+(-14)=?$

4. Solve the system by elimination. $\begin{aligned} 3x - y &= 21 \\ 2x + y &= 4 \end{aligned}$

5. Simplify. $\frac{15x^2yz^4}{20x^2y^2z^2}$

6. Write an expression for *the difference of a number and eleven.*

7. $0.72 - 0.5634 = ?$

8. Put these integers in increasing order. –29, –3, –17, –25

9. Write 0.0067 in scientific notation.

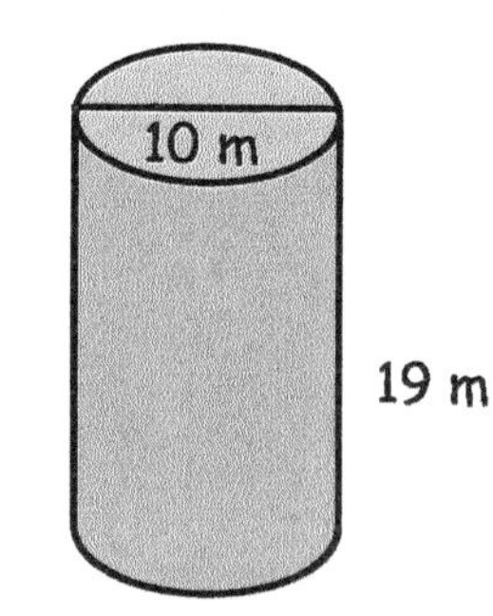

10. Calculate the volume of the cylinder.

11. Solve for x. $3(x-8)=5x-4$

12. Solve for x. $\frac{x}{3} = \frac{8}{24}$

13. $7 - a \leq 12$ Graph the solution on a number line.

14. The Fahrenheit boiling point of water is ________.

15. $\begin{pmatrix} -4 & 13 \\ 8 & 1 \end{pmatrix} - \begin{pmatrix} 4 & -5 \\ 10 & -14 \end{pmatrix} = ?$

16. Solve for p. $|p-9|=3$

17. $6{,}374{,}219 + 7{,}998{,}366 = ?$

18. Solve for a. $5(a-1)=35$

19. What is the value of c in $c+21=-96$?

20. The price of a wide-screen TV went from \$3,500 to \$4,000. By what percent did it increase? (Hint: Round to the nearest whole number.)

1.	2.	3.	4.
5.	6.	7.	8.
9.	10.	11.	12.
13.	14.	15.	16.
17.	18.	19.	20.

Lesson #26

1. Solve the system of equations using any method you choose. $y = x + 2$, $y = -2x + 3$

2. $-56 - 38 = ?$

3. Simplify. $4(3n + 4m + 4) - 2(2n + 6)$

4. Write the slope-intercept form of a linear equation.

5. Write the equation for the line through point (3, – 4) with a slope of 6.

6. What is the probability of getting heads, tails, and heads on 3 flips of a coin?

7. Solve for x. $x - 121 = 216$

8. $\dfrac{-125}{-5} = ?$

9. Which is greater, 0.75 or $\dfrac{4}{5}$?

10. Write 2.0075×10^6 in standard form.

11. $0.07 \times 0.003 = ?$

12. Solve for a. $\dfrac{4a + 4}{3} = 12$

13. What is the value of x in $10x - 7 = 17 + 2x$?

14. How many feet are in 9 yards?

15. A nine-sided figure is called a(n) ________.

16. $30 \div 6 + 4 \cdot 3 - 6 - 1 = ?$

17. $52\frac{1}{8} - 37\frac{5}{8} = ?$

18. Solve for a. $\dfrac{1}{10}a + 15 = 27$

19. Evaluate $5x - 2y$ when $x = 3$ and $y = 2$.

20. The long jump record at your school is 17 feet $8\frac{1}{4}$ inches. Your best long jump so far is 15 feet $11\frac{3}{4}$ inches. How much farther do you need to jump to match the school record?

1.	2.	3.	4.
5.	6.	7.	8.
9.	10.	11.	12.
13.	14.	15.	16.
17.	18.	19.	20.

Lesson #27

1. A male orangutan is about $3\frac{2}{5}$ feet tall. A female is about $3\frac{3}{7}$ feet tall. Which orangutan is shorter?
2. $-5(12)=?$
3. When $a=5$ and $b=9$, what is the value of $\frac{ab}{3}+a$?
4. Solve for a. $23a+9=66+4a$
5. $24.315 \div 0.3 = ?$
6. Solve for b. $b+29=-74$
7. Solve the system using the method of your choice. $\begin{aligned} y&=4 \\ 3x-y&=5 \end{aligned}$
8. Find the values of y in the equation $y=12-x$, when $x=\{-2, 7, 0\}$.
9. How many yards are in 3 miles?
10. Solve for x. $\frac{3}{5}x=45$
11. Write the formula for finding the volume of a cylinder.
12. $40+5\cdot3-6\cdot2+8\div2=?$
13. Represent *the sum of a number and ten* as an algebraic phrase.
14. Find $\frac{5}{7}$ of 49.
15. $\frac{8}{10}\div\frac{2}{5}=?$
16. Solve for x. $-4x-2<8$ (Give the answer as an improper fraction.)

17. True or false: The slope of a vertical line is undefined.
18. Solve for x. $\frac{x}{14}+2=10$
19. Write $\frac{7}{50}$ as a decimal and as a percent.
20. Find the slope of a line passing through points (9, –2) and (3, 4).

1.	2.	3.	4.
5.	6.	7.	8.
9.	10.	11.	12.
13.	14.	15.	16.
17.	18.	19.	20.

Lesson #28

1. How many pints are in 6 quarts?

2. $12 > -3x$ and $-2x > -12$ Graph the solution on a number line.

3. A seven-sided shape is called a(n) ________.

4. Write the equation for the line through point (4, –8) with a slope of 3.

5. What is the measure of the angle that is complementary to the angle shown?

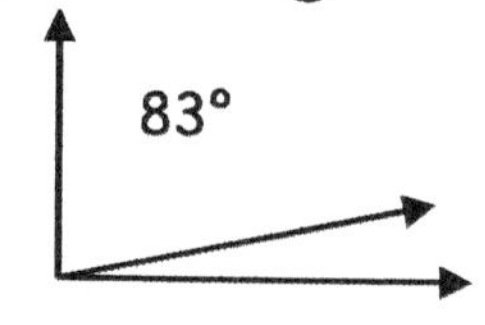

6. Graph the inequality on a number line. $11w \leq -22$

7. Solve for x. $\frac{1}{6}x - 15 = -25$

8. Determine the area of a triangle with a base of 15 cm and a height of 4 cm.

9. Write 0.00083 in scientific notation.

10. A gel pen, normally \$1.75, has been reduced to \$1.25. By what percent has the price of the pen been reduced? (Round to the nearest whole number.)

11. $132 - (-58) = ?$

12. What number is 40% of 25?

13. $9 - 3\frac{4}{5} = ?$

14. Solve for x. $9x + 6 = -15 + 6x$

15. $1.25 \times 0.4 = ?$

16. Two angles whose measures add up to 180° are ________.

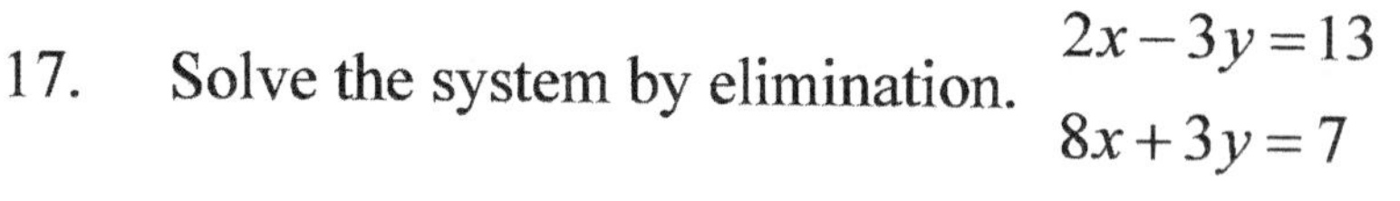

17. Solve the system by elimination. $\begin{matrix} 2x - 3y = 13 \\ 8x + 3y = 7 \end{matrix}$

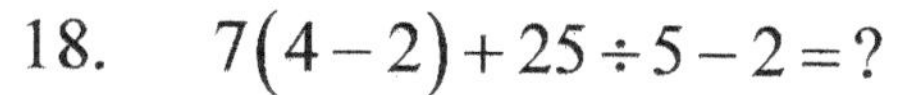

18. $7(4 - 2) + 25 \div 5 - 2 = ?$

19. Find the average of 75, 65 and 85.

20. Simplify. $\frac{3x^2y^2}{9xy^3z}$

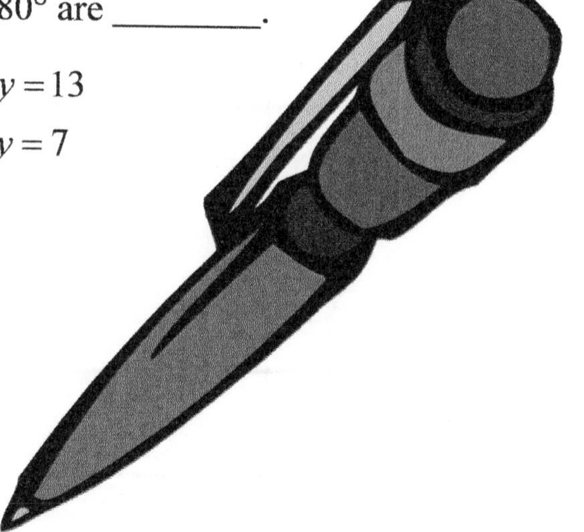

1.	2.	3.	4.
5.	6.	7.	8.
9.	10.	11.	12.
13.	14.	15.	16.
17.	18.	19.	20.

Lesson #29

1. $-426+(-266)=?$

2. Use any method to solve the system. $\begin{aligned}-5x+2y&=14\\-3x+y&=-2\end{aligned}$

3. $49.77 + 36.8 = ?$

4. Find the values for y in the equation $y=x-4$ when $x = \{0, 3, 6\}$.

5. Solve the inequality and graph the solution on a number line. $\frac{3}{5}x>6$

6. $17-9\frac{4}{9}=?$

7. How many feet are in 5 miles?

8. A small can of coffee used to weigh 18 ounces, but now weighs 12 ounces. By what percent has the weight of the can changed? (Round to the nearest whole number.)

9. Write 3.24×10^5 in standard form.

10. What is the volume of a rectangular prism if its length is 19 cm, its width is 8 cm, and its height is 4 cm?

11. Solve for x. $\frac{x}{15}+3=12$

12. Write 0.42 as a percent and as a reduced fraction.

13. $\begin{pmatrix}6 & 0 & -7\\2 & -9 & 3\end{pmatrix}+\begin{pmatrix}5 & -5 & -6\\3 & -2 & 7\end{pmatrix}=?$

14. $\frac{-240}{-80}=?$

15. $32\div4+3\cdot6-12\div4+2=?$

16. $-9^3 = ?$

17. Evaluate $\frac{abc}{3}+ab$ if $a=2$, $b=3$, and $c=4$.

18. Solve for x. $|x|+5=11$

19. Solve for x. $\frac{1}{5}x+13=22$

20. What value of a makes the equation true? $15a+6=4a+17$

1.	2.	3.	4.
5.	6.	7.	8.
9.	10.	11.	12.
13.	14.	15.	16.
17.	18.	19.	20.

Lesson #30

1. Find the value of b. $\frac{3}{8}b = 9$

2. $-8(4)(-3) = ?$

3. $\frac{4}{5} \times \frac{15}{16} = ?$

4. Use elimination to solve the system. $x - y = 12$; $x + y = 22$

5. A jar contains five blue balls, three yellow, six green, and two purple balls. If you choose two balls, what is the P(purple and blue) with replacement? What is the P(green and blue) without replacement?

6. The area of a square is 169 mm^2. What is the measure of each side?

7. Solve for x. $x - 22 = -56$

8. Round 82.463 to the nearest tenth.

9. Solve for x. $3x - 9 = 15$

10. $15 + 3[2(6+2) - 3] = ?$

11. How many pints are in 5 quarts?

12. Find the solution set. $7y + 6 \leq -8$

13. Find the median and the range of 12, 25, 16, 32 and 50.

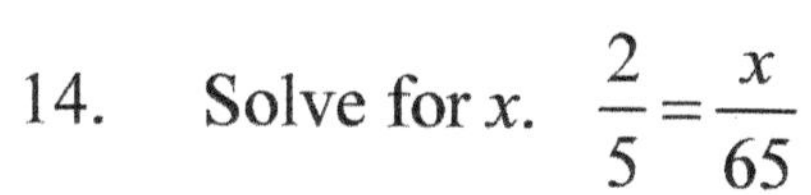

14. Solve for x. $\frac{2}{5} = \frac{x}{65}$

15. What is the formula for finding the surface area of a rectangular prism?

16. Find the missing measurement of angle, h.

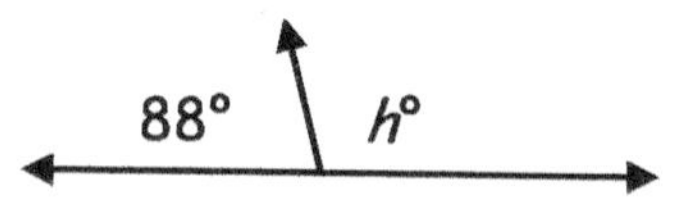

17. A ten-sided shape is called a(n) ________.

18. Write 2,000,000 in scientific notation.

19. What value of x makes the equation true? $2x - 10 = 7x$

20. A line passing through the points (–5, –7) and (–1, 3) has what slope?

1.	2.	3.	4.
5.	6.	7.	8.
9.	10.	11.	12.
13.	14.	15.	16.
17.	18.	19.	20.

Lesson #31

1. Solve the inequality for y. $y+3<16$

2. $99+(-62)=?$

3. Use any method to solve the system. $\begin{aligned} x-y&=7 \\ 3x+2y&=6 \end{aligned}$

4. Solve for b. $b-76=-102$

5. Two angles whose measures add up to 90° are ________ angles.

6. $n-5\geq-13$ Graph the solution on a number line.

7. Write an equation for the line through point (8, 5) having a slope of $\frac{1}{2}$.

8. Find the percent of change from 7 feet to 2 feet. (Round to the nearest whole number.)

9. 700 centimeters are how many meters?

10. Solve for x. $\frac{x}{4}-8=16$

11. Find the volume of the cylinder.

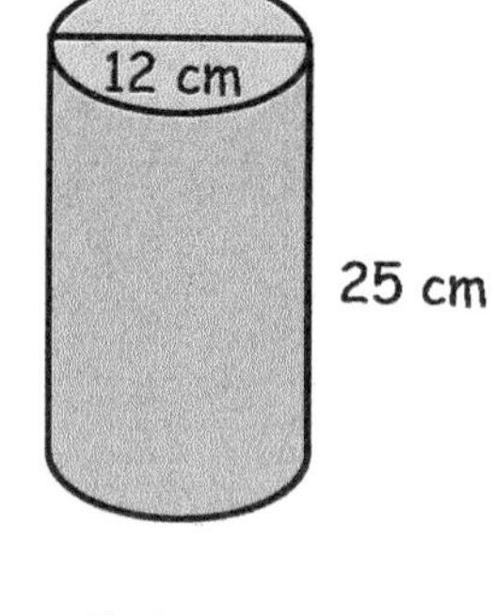

12. $8\cdot2+15\div3+2^3=?$

13. Write the formula for finding the area of a parallelogram.

14. Find the GCF of $12x^3y^2$ and $16x^2y^5$.

15. What is the probability of getting heads three times in a row on 3 flips of a coin?

16. Simplify. $7(4a+2b-5)-3(2a+4)$

17. Write 0.0058 in scientific notation.

18. 60% of what number is 18?

19. Solve for c. $3c-12=-9$

20. $14\frac{1}{7}+23\frac{2}{3}=?$

1.	2.	3.	4.
5.	6.	7.	8.
9.	10.	11.	12.
13.	14.	15.	16.
17.	18.	19.	20.

Lesson #32

1. Solve the system by using any method. $x + y = 19$ $\quad x - y = -7$

2. The circumference of a circle with a diameter of 12 mm is __________.

3. Solve for x. $\frac{5}{9} = \frac{x}{108}$

4. Your friend factored $18x^2y$ as $3 \cdot 6 \cdot x \cdot y$. Correct your friend's error and write it correctly.

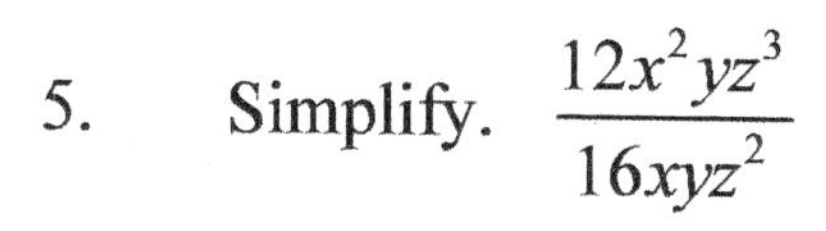

5. Simplify. $\frac{12x^2yz^3}{16xyz^2}$

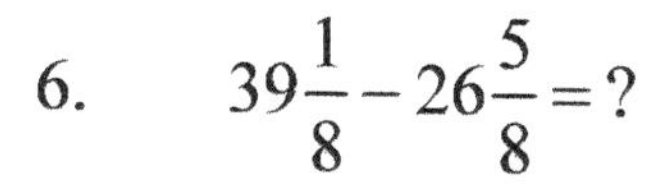

6. $39\frac{1}{8} - 26\frac{5}{8} = ?$

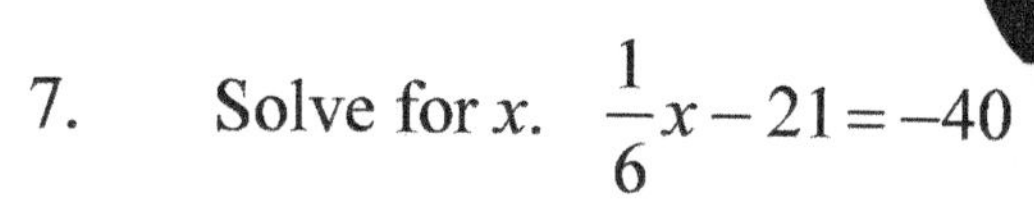

7. Solve for x. $\frac{1}{6}x - 21 = -40$

8. $\begin{pmatrix} 5 & 0 & -2 \\ 3 & -7 & -4 \end{pmatrix} + \begin{pmatrix} 2 & -9 & -3 \\ -4 & 6 & -2 \end{pmatrix} = ?$

9. Solve for x. $\frac{x}{15} - 12 = 19$

10. When $x = \{0, -3, 2\}$ in $y = 2x + 2$, what are the corresponding values of y?

11. Put these decimals in decreasing order. 2.04 2.47 2.047 2.0

12. What value of h makes the equation true? $h + 28 = 64$

13. Write 54% as a decimal and as a reduced fraction.

14. $(-2)^7 = ?$

15. Write 1.234×10^4 in standard form.

16. Solve the equation for a. $12a - 4 = 14 + 9a$

17. Write the slope and the y-intercept for the line whose equation is $y = 5x - 9$.

18. Solve for h. $6h - 11 = 13$

19. How many degrees are in a circle?

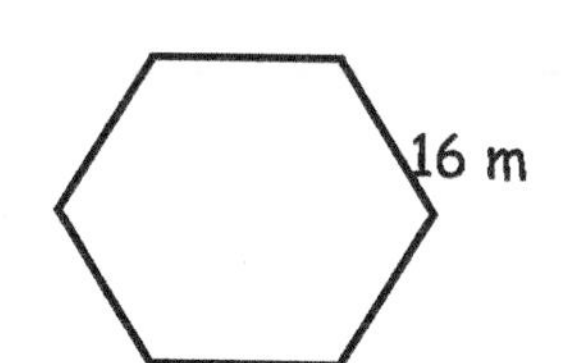

20. Find the perimeter of the hexagon.

1.	2.	3.	4.
5.	6.	7.	8.
9.	10.	11.	12.
13.	14.	15.	16.
17.	18.	19.	20.

Lesson #33

1. The slope of a horizontal line is __________.

2. Solve for m. $5m+4=8m-2$

3. $-362+(-198)=?$

4. Write 780,000,000 in scientific notation.

5. Solve for x. $5x+5=10$

6. Find the slope of the line passing through points (1, 2) and (6, 1).

7. What is the value of x in $\frac{x}{8}-7=15$?

8. Find the LCM of $10x^2y^2z^2$ and $15xy^3z^4$.

9. Find the area of the trapezoid shown to the right.

26 m
22 m
18 m

10. What number is 15% of 96?

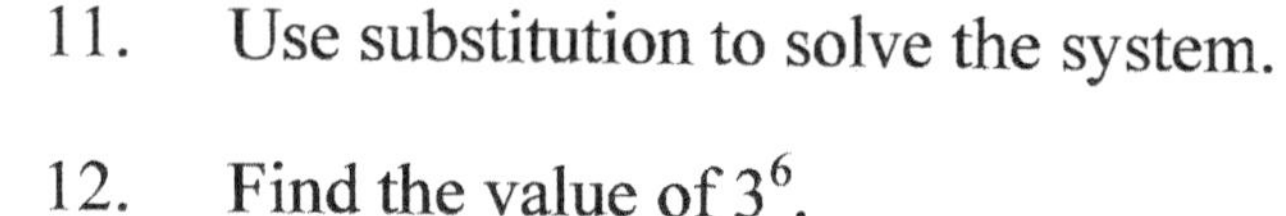
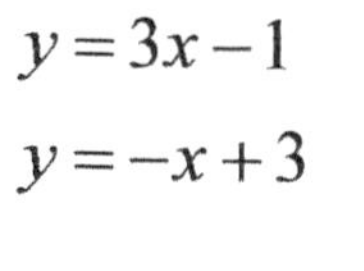

11. Use substitution to solve the system. $y=3x-1$, $y=-x+3$

12. Find the value of 3^6.

13. Solve for a. $\frac{3}{5}a-\frac{2}{5}a+3=12$

14. $[54\div 6+(3\cdot 3)]\div 2+3^2=?$

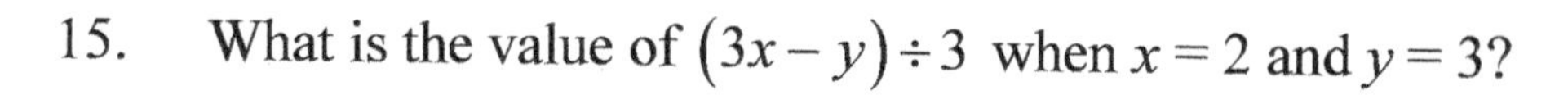

15. What is the value of $(3x-y)\div 3$ when $x=2$ and $y=3$?

16. A greeting card was $1\frac{2}{5}\times 3\frac{1}{2}$ inches. Find the area of the greeting card.

17. $12n<3n+27$ Solve the inequality and graph in on a number line.

18. An earthworm grew from 2.3 cm to 2.8 cm in length. By what percent did the worm grow? (Round to the nearest whole number.)

19. Solve for x. $-8x=112$

20. Which is greater, $\frac{4}{5}$ or 75%?

1.	2.	3.	4.
5.	6.	7.	8.
9.	10.	11.	12.
13.	14.	15.	16.
17.	18.	19.	20.

Lesson #34

1. Use elimination to solve the system. $\begin{array}{l} x+y=10 \\ x-y=2 \end{array}$

2. $2x+2x>4$ Solve the inequality and graph the solution on a number line.

3. Write the slope-intercept form of a linear equation.

4. Put these integers in increasing order. $-13,\ 0,\ 12,\ -5,\ 7$

5. Solve for x. $-3<x+2<7$

6. $12-7\frac{5}{6}=?$

7. Write 0.00125 in scientific notation.

8. Simplify. $8(4x-2y)+3(2x-2y)$

9. $34-(-17)=?$

10. $\frac{3x-2}{5}=-7$

11. Find $\frac{4}{7}$ of 63.

12. $60+4\cdot5-12\div2+3=?$

13. Write the values for y in the equation $y=x+3$ when $x=\{0, -5, 6\}$.

14. Solve for x. $4x-7=2x+7$

15. What is the value of b in $b-25=86$?

16. Hopewell Middle School had 246 students. Two-thirds of the students were on the honor roll during the first quarter. How many of the students were not on the honor roll?

17. Find the slope of a line passing through points $(0, -4)$ and $(2, -2)$.

18. Translate into an algebraic expression: *Nine times a number decreased by eleven.*

19. $4.816\div0.4=?$

20. Write the formula for finding the area of a circle.

1.	2.	3.	4.
5.	6.	7.	8.
9.	10.	11.	12.
13.	14.	15.	16.
17.	18.	19.	20.

Lesson #35

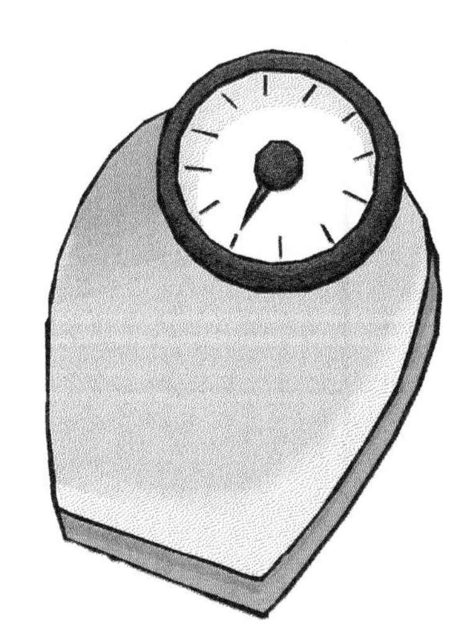

1. Find the value of x. $\frac{11x-3}{6}=5$

2. Solve for x. $-7x-8>16+5x$

3. Which is greater, $\frac{1}{4}$ or 0.35?

4. Write 3.445×10^5 in standard form.

5. Solve for x. $\frac{x}{5}=-60$

6. Solve for a. $17+a=7a-13$

7. What value of a makes the equation true? $3(2a+4)=24$

8. The average weight of 3 adults is 235 pounds. One of the adults weighs 226 pounds. A 2nd adult weighs 195 pounds. What is the weight of the 3rd adult?

9. $-92+(-58)=?$

10. How many centimeters are 5 meters?

11. $(-2)(-3)(-9)=?$

12. $\frac{4}{9}\times\frac{18}{24}=?$

13. A triangle with no congruent sides is __________.

14. The price of a stereo system is reduced from $750 to $600. By what percent has the price been reduced?

15. Write the equation of the line passing through points (7, 3) and (2, 2).

16. $1.3 \times 2.6 = ?$

17. Solve for x. $7x-9=19$

18. $20-2[3+12\div 4]=?$

19. Evaluate $x+\frac{y}{2}+xy$ when $x=4$ and $y=6$.

20. Use substitution to solve the system of equations. $y=-2x+3$ $y=x-6$

1.	2.	3.	4.
5.	6.	7.	8.
9.	10.	11.	12.
13.	14.	15.	16.
17.	18.	19.	20.

Lesson #36

1. Solve for x. $4x+10=2x-22$

2. Solve for x. $\frac{5}{9}x=45$

3. How many cups are in 4 pints?

4. What is the missing measure, y?

5. Write 5.97×10^{-4} in standard form.

6. What is the value of x in $\left(\frac{1}{8}x+14=20\right)$?

7. $7-a \le 12$ Graph the solution to the inequality on a number line.

8. $0.7-0.3662=?$

9. Solve for x. $\frac{4}{5}=\frac{x}{65}$

10. $\frac{5}{6} \bigcirc \frac{7}{8}$

11. Find the surface area of this cube.

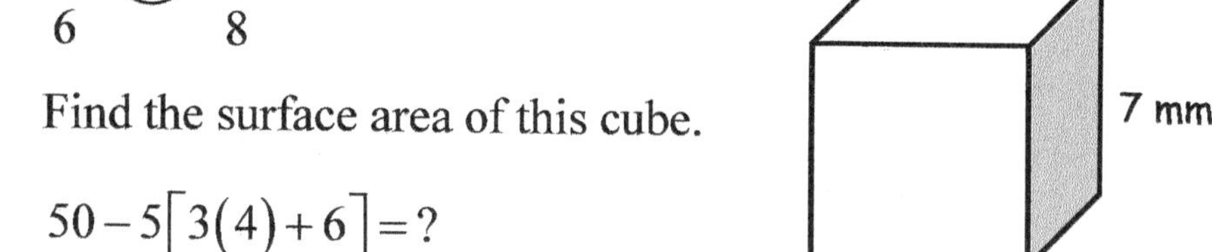

12. $50-5[3(4)+6]=?$

13. Solve for a. $\frac{5a+2}{8}=4$

14. Use any method to solve the system. $x+y=19$; $10x-7y=20$

15. Write an equation for the line through points (2, 9) and (7, 10).

16. A rectangle is 19 mm long and 8 mm wide. What is its area?

17. A survey of 60 people finds that 18 of them sleep eight hours each night. What percent of the group sleeps eight hours a night?

18. 30% of what number is 24?

19. $700{,}000-259{,}667=?$

20. Simplify. $\frac{25a^3b^2}{30abc^4}$

1.	2.	3.	4.
5.	6.	7.	8.
9.	10.	11.	12.
13.	14.	15.	16.
17.	18.	19.	20.

Lesson #37

1. $4^{-3} = ?$

2. Find and graph the solution to $5 - 2n \le 3 - n$.

3. Solve for x. $\frac{1}{3}x = 15$

4. Write the equation for the line through points (25, 100) and (15, 120).

5. What value of x makes the equation true? $13x - 6 = 5x + 26$

6. How many quarts are in 12 gallons?

7. Write 0.0097 in scientific notation.

8. Two angles whose measures add up to 180° are ___________ angles.

9. Solve for x. $\frac{x}{17} - 6 = 24$

10. $14 - 8\frac{3}{8} = ?$

11. In the equation $y = 2x - 4$, when $x = \{0, 4, -2\}$, what are the corresponding values of y?

12. Solve the system using the method of your choice. $\begin{array}{l} y = 2x \\ 7x - y = 35 \end{array}$

13. On 2 consecutive rolls of a die, what is the probability of rolling a 2 and a 4?

14. Solve for x. $4(x - 8) = -12$

15. Find the median, the mode, and the range of 28, 42, 66, 19 and 28.

16. $129(-3) = ?$

17. Evaluate $5a - 2b$ when $a = 3$ and $b = 2$.

18. Calculate the perimeter of a decagon if the sides measure 15 inches each.

19. Write an algebraic expression: *One-third of a number decreased by four.*

20. The price of your stock began at \$35. The price fluctuated over a five month period: down \$3, down \$7, up \$5, down \$1, and up \$9. What was the value of your stock at the end of the five months?

1.	2.	3.	4.
5.	6.	7.	8.
9.	10.	11.	12.
13.	14.	15.	16.
17.	18.	19.	20.

Lesson #38

1. Solve and graph the inequality on a number line. $-6m \geq 36$

2. A line passes through point (–5, 2) with a slope of 0. Write its equation.

3. Solve for x. $\frac{x}{13} = -9$

4. 80% of what number is 72?

5. Use any method to solve the system. $\begin{aligned} x + 5 &= y \\ x + y &= 1 \end{aligned}$

6. $-4^{-3} = ?$

7. Solve for x. $x - 22 = 84$

8. Water freezes at ______ °F.

9. Write the slope and the y-intercept for a line whose equation is $y = \frac{4}{5} + 7$.

10. Find the volume of the cylinder.

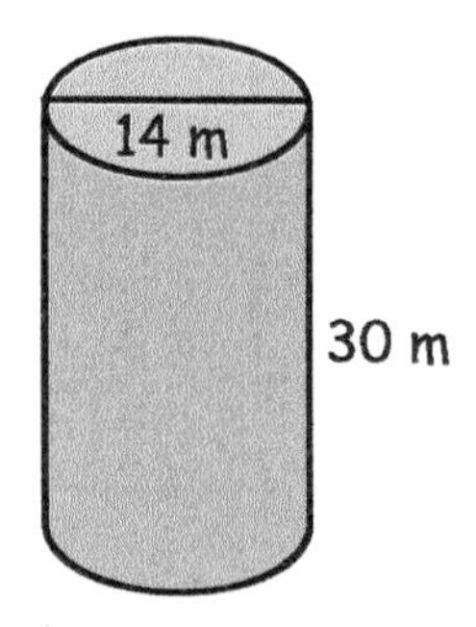

11. $-2.36^0 = ?$

12. Write 400,000,000 in scientific notation.

13. Solve the proportion for x. $\frac{5}{6} = \frac{x}{24}$

14. What values of f make the equation true? $f - 10 \leq 16$

15. $16 \div 4 + 8[2 + 12 \div 6] = ?$

16. Solve for y. $3y + 25 = -11$

17. How many grams are in 5 kilograms?

18. In the equation $21 - x = 6x$, what is the value of x?

19. The slope of a vertical line is ___________.

20. $28\frac{1}{3} - 16\frac{2}{5} = ?$

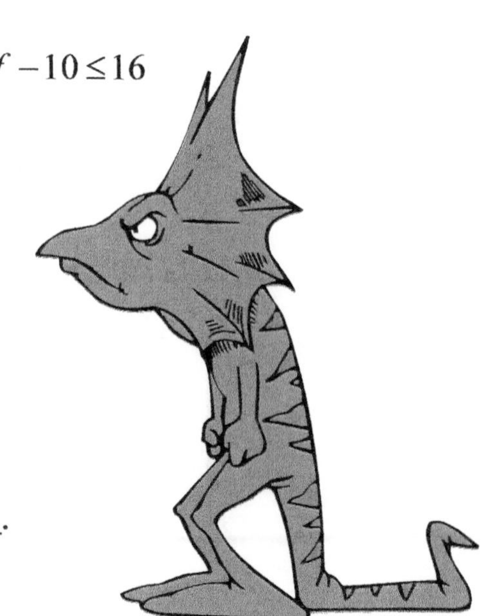

1.	2.	3.	4.
5.	6.	7.	8.
9.	10.	11.	12.
13.	14.	15.	16.
17.	18.	19.	20.

Lesson #39

1. Write each using only positive exponents. a) $\dfrac{7ab^{-2}}{3w}$ b) $\dfrac{1}{x^{-7}}$ c) $\dfrac{5^{-2}}{a}$

2. Solve for x. $\dfrac{2}{3}x = -18$

3. $3.2 - 1.975 = ?$

4. Simplify. $5(2x + 3y + 4) + 6x - 9$

5. As the sun rose, the length of Tara's shadow went from 40 inches to 25 inches. By what percent did the length of her shadow decrease?

6. $\begin{pmatrix} 3 & -9 \\ 2 & 0 \end{pmatrix} - \begin{pmatrix} -3 & -6 \\ 1 & -5 \end{pmatrix} = ?$

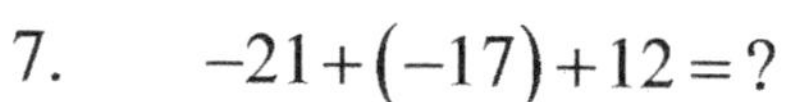

7. $-21 + (-17) + 12 = ?$

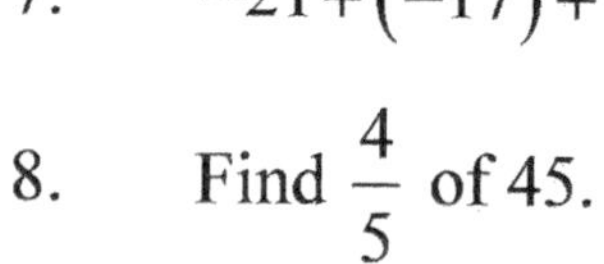

8. Find $\dfrac{4}{5}$ of 45.

9. In the equation $7x - 12 = 16$, what is the value of x?

10. Determine the equation of the line through point (1, 2) with a slope of –3.

11. $0.4 \div 0.08 = ?$

12. Write $\dfrac{3}{25}$ as a decimal and as a percent.

13. Write 4.2×10^4 in standard form.

14. $(-4)^2 = ?$

15. Use the method of your choice to solve the system. $\begin{matrix} y = x + 2 \\ y = -2x + 3 \end{matrix}$

16. If a circle's diameter is 48 millimeters, what is the length of its radius?

17. Write the formula for finding the area of a parallelogram.

18. $14\dfrac{1}{5} + 13\dfrac{1}{3} = ?$

19. $3b + 12 > 21 - 2b$ Determine the solution to the inequality and graph it.

20. Evaluate $\dfrac{2a}{b} - c$ when $a = 15$, $b = 5$, and $c = 2$.

1.	2.	3.	4.
5.	6.	7.	8.
9.	10.	11.	12.
13.	14.	15.	16.
17.	18.	19.	20.

Lesson #40

1. Solve the inequality and graph the solution on a number line. $-4 < h+1 < 7$

2. $2 \cdot 4[5+(6-2)+3]=?$

3. $-5(-6)(3)=?$

4. Write an algebraic phrase that means *the product of a number and seven.*

5. Solve for x. $\frac{x}{9}-7=16$

6. Jeffery ran 13 miles in 1 hour 32 minutes. About how long did it take him to run one mile?

7. What value of x will make this equation true? $7x-19=30$

8. What is the probability of rolling a prime number on one roll of a die?

9. $\frac{8}{9} \times \frac{18}{24}=?$

10. A change from 15 inches to 5 inches represents a decrease of what percent?

11. When $a=4$ and $b=6$, what is the value of $7a-b+2$?

12. Rewrite each using only positive exponents. a) $5ac^{-5}$ b) $x^{-5}y^{-6}$

13. $151^0=?$

14. Solve for x. $|x|+4=15$

15. What value of b makes the two sides of the equation equal? $b-22=-81$

16. Evaluate $a^{-b}b$ when $a=3$ and $b=2$.

17. Solve the system using any method. $\begin{aligned} y-3&=x \\ x+2y&=3 \end{aligned}$

18. Solve for a. $20a-14=6a+28$

19. $375{,}896+299{,}853=?$

20. Write an equation for the line through points (7, 3) and (2, 2).

1.	2.	3.	4.
5.	6.	7.	8.
9.	10.	11.	12.
13.	14.	15.	16.
17.	18.	19.	20.

Lesson #41

1. Evaluate b^{-a} when $a = 3$ and $b = 2$.

2. $\begin{pmatrix} 9 & -3 \\ -6 & 2 \end{pmatrix} - \begin{pmatrix} 7 & -4 \\ 0 & 3 \end{pmatrix} = ?$

3. $-69 + (-91) = ?$

4. Solve for x. $x + 96 = -128$

5. Find the area of the triangle.

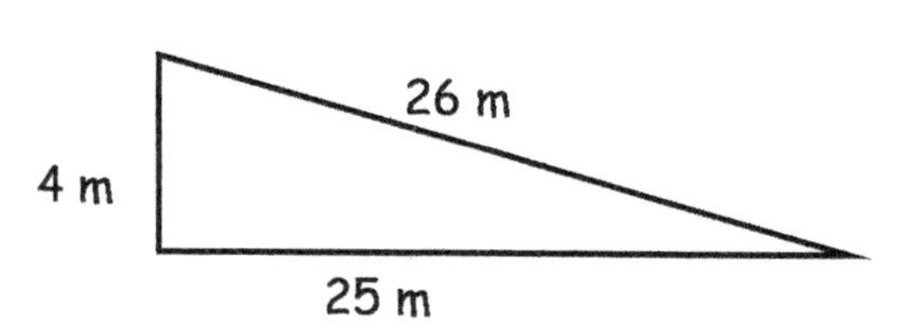

6. Simplify. $\dfrac{3x^2y^3}{18xy^2z}$

7. Write the slope-intercept form of a linear equation.

8. Rewrite the expression so that it contains only positive exponents. $\dfrac{1}{a^{-3}b^3}$

9. 20% of what number is 18?

10. Write 85% as a decimal and as a reduced fraction.

11. Translate the sentence into an algebraic expression: *The difference of a number and twelve.*

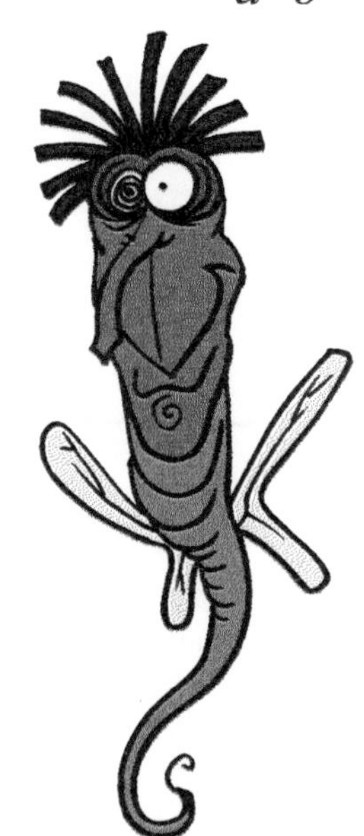

12. $3.624 \div 0.06 = ?$

13. $4 + h > 3$ Solve the inequality and graph the solution on a number line.

14. Solve for b. $\dfrac{1}{3}b - 12 = 9$

15. Find the value of a. $11a + 45 = 2a$

16. Write 0.00000213 in scientific notation.

17. $86.24 + 19.893 = ?$

18. How many feet are in 8 yards?

19. Solve using substitution. $y = 5x - 8$, $5y = 2x + 6$

20. Find the coordinates of points A and B.

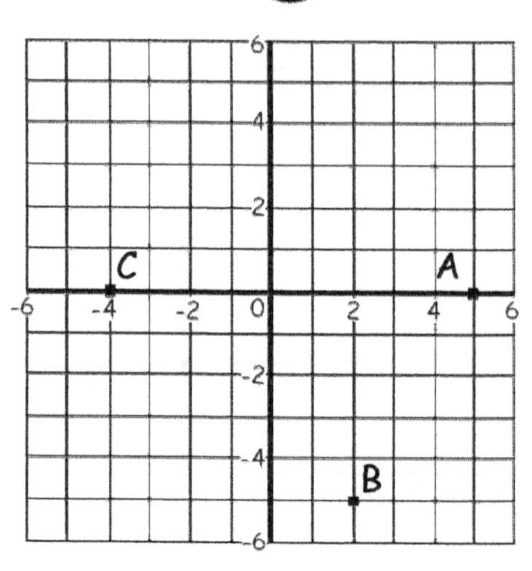

1.	2.	3.	4.
5.	6.	7.	8.
9.	10.	11.	12.
13.	14.	15.	16.
17.	18.	19.	20.

Lesson #42

1. Write the decimal number in scientific notation. 0.00325

2. Rewrite without using negative exponents. $\frac{8a^{-5}}{c^{-3}d^{3}}$

3. $\begin{pmatrix} 2 & 9 & 5 \\ 3 & 6 & 1 \end{pmatrix} + \begin{pmatrix} 5 & -3 & 9 \\ 2 & 4 & -6 \end{pmatrix} = ?$

4. Factor. $24a^3bc^2$

5. How many years are 5 decades?

6. Write the formula for finding the surface area of a rectangular prism.

7. Change from scientific notation to standard notation. 7.042×10^9

8. Solve for x. $25x + 12 = 9x - 20$

9. Find the missing measurement, x.

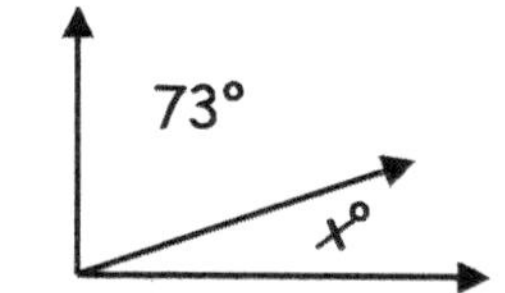

10. $93 - (-41) = ?$

11. Find the LCM of $15x^3y^2$ and $30x^4y^3z$.

12. Solve for v. $|v - 3| \geq 4$

13. What value of x makes the equation true? $\frac{x}{12} - 8 = 14$

14. A line through point (0, 3) has a slope of 1. Write the equation for this line.

15. Solve for a. $-3a = -39$

16. $42 - 30\frac{1}{8} = ?$

17. What is the value of x in $\frac{6}{x} = \frac{3}{5}$?

18. After moving to a new house, Sami's ride to school changed from 6 km to 6.5 km. By what percent did it increase? (Round to the nearest whole number.)

19. 35% of 15 is what number?

20. Which is greater, 0.09 or $\frac{2}{25}$?

1.	2.	3.	4.
5.	6.	7.	8.
9.	10.	11.	12.
13.	14.	15.	16.
17.	18.	19.	20.

Lesson #43

1. What values of b make this inequality true? $9 - b \leq 4$

2. Solve for x. $\frac{1}{11}x - 9 = 1$

3. Write 0.00092 in scientific notation.

4. A jar of gumballs contains 5 blue, 3 yellow, 6 green, and 2 purple gumballs. If two gumballs are selected from the jar, what is the P(green and yellow) with replacement?

5. Solve using any method. $4x - 5y = 9$; $-2x - y = -29$

6. Rewrite using only positive exponents: $5m^5m^{-7}$.

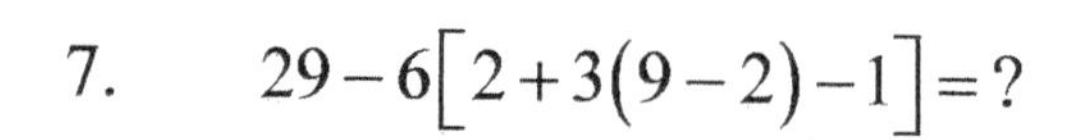

7. $29 - 6[2 + 3(9 - 2) - 1] = ?$

8. $13 + 15 + (-10) = ?$

9. $-8^3 = ?$

10. Write 8.05×10^6 in standard notation.

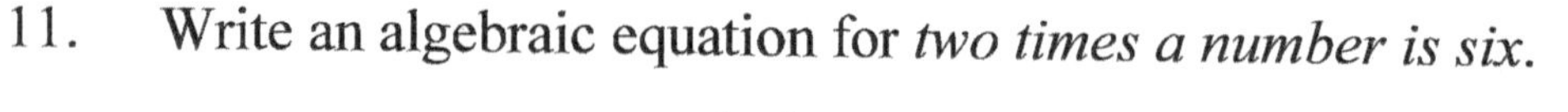

11. Write an algebraic equation for *two times a number is six.*

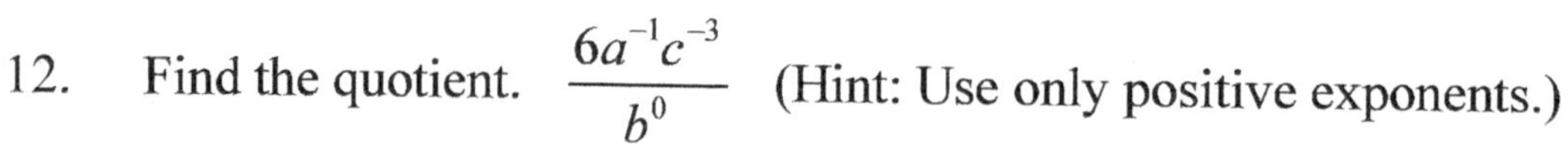

12. Find the quotient. $\frac{6a^{-1}c^{-3}}{b^0}$ (Hint: Use only positive exponents.)

13. $3.4 \times 0.7 = ?$

14. Evaluate the expression $a + 4b - 2$ when $a = 9$ and $b = 2$.

15. How many meters are 300 centimeters?

16. $|42| = ?$

17. $\sqrt{49} + \sqrt{144} - 2^2$

18. Susan bought a new couch for $950. The tax was 7%. What was the total cost of the couch?

19. Solve the equation for b. $7b - 4 = 2b + 6$

20. Write the ratio *3:5* in two other ways.

1.	2.	3.	4.
5.	6.	7.	8.
9.	10.	11.	12.
13.	14.	15.	16.
17.	18.	19.	20.

Lesson #44

1. Rewrite the expression using only positive exponents. $\dfrac{c^4}{x^2y^{-1}}$

2. Write the formula that is used to find the area of a trapezoid.

3. Write 83.5×10^{-6} in correct scientific notation.

4. Simplify. $\dfrac{12a^2b}{18abc^2}$

5. Solve for h. $h + 44 = 61$

6. Solve for x. $5x + 6 = -19$

7. How many cups are in 8 pints?

8. Write the formula for finding the volume of a cylinder.

9. Solve using any method. $x + y = 4$, $y = 7x + 4$

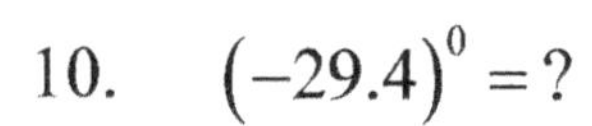

10. $(-29.4)^0 = ?$

11. $\dfrac{s}{6} \leq 3$ Solve the inequality and graph its solution.

12. Write $\dfrac{7}{25}$ as a decimal and as a percent.

13. Find the slope of the line passing through points (1, 6) and (9, – 4).

14. What number is 25% of 80?

15. $35\frac{2}{7} - 17\frac{6}{7} = ?$

16. What value of a will make the equation true? $14a - 7 = 7a$

17. $60 + 3 \cdot 5 - 2[10 \div 2] = ?$

18. Solve for y when $x = \{-1, 0, 5\}$. $y = x - 3$

19. Solve for x. $\dfrac{x}{8} + 5 = 11$

20. $900{,}000 - 623{,}175 = ?$

1.	2.	3.	4.
5.	6.	7.	8.
9.	10.	11.	12.
13.	14.	15.	16.
17.	18.	19.	20.

Lesson #45

1. Rewrite without using negative exponents. $\dfrac{mn^{-4}}{pq^{-2}}$

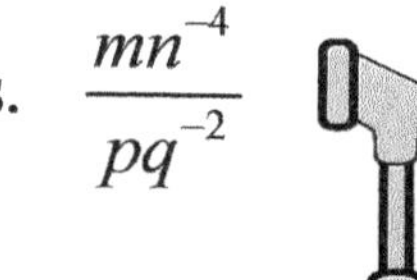

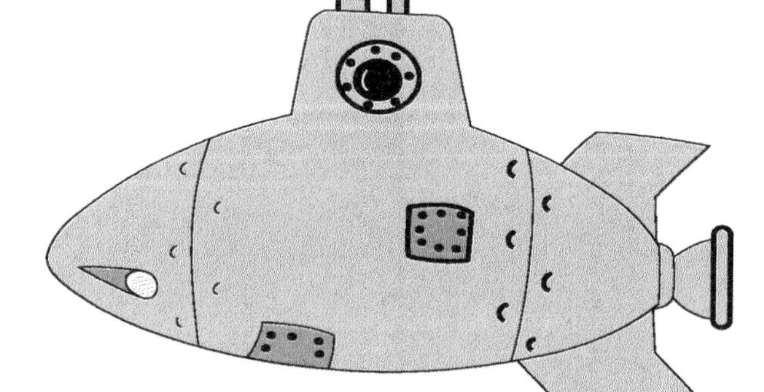

2. $4.69 \times 10^{-6} = ?$

3. $-218 - (-89) = ?$

4. $\begin{pmatrix} 2 & 9 & 5 \\ -6 & 0 & -4 \end{pmatrix} - \begin{pmatrix} -1 & 4 & -2 \\ -2 & -1 & 2 \end{pmatrix} = ?$

5. Solve for x. $\dfrac{x}{19} + 12 = 15$

6. What is the equation of the line passing through points (3, –3) and (–3, 1)?

7. How many yards are in 4 miles?

8. Solve to find the value of x. $7x - 3 = 4x + 6$

9. What percent of 90 is 36?

10. Solve the system using substitution. $\begin{aligned} 3x - y &= 17 \\ 2x + y &= 8 \end{aligned}$

11. Solve for b. $\dfrac{3}{4}b = 12$

12. A triangle with a base of 42 mm and a height of 4 mm has what area?

13. Evaluate t^{-m} when $t = -3$ and $m = 2$.

14. What value of x makes the equation true? $x - 68 = 142$

15. $92 + 3 \cdot 2 - 24 \div 6 + 2^2 = ?$

16. Find the surface area of the cube.

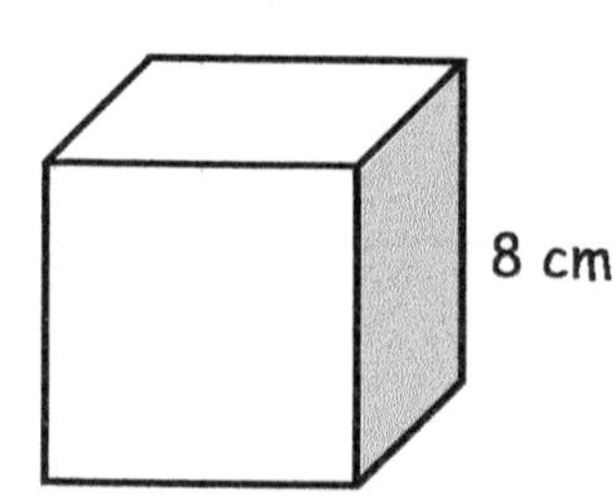

17. $86 - 49\dfrac{5}{6} = ?$

18. Write the decimal in scientific notation. 0.00063

19. $9 + c \leq 17$ Graph the solution on a number line.

20. $9.206 \div 0.2 = ?$

1.	2.	3.	4.
5.	6.	7.	8.
9.	10.	11.	12.
13.	14.	15.	16.
17.	18.	19.	20.

Lesson #46

1. Put these integers in decreasing order. –92, –46, –3, 0, –55
2. How many sides are in a heptagon?
3. Solve the inequality for w. $8 - w < 8$
4. What is the value of b^{-a} when $a = 3$ and $b = 2$?
5. $1.097 \times 10^6 = ?$
6. Find the LCM of $4a^2b^3c^6$ and $18a^5b^2c^4$.
7. $27 + 56 + (-33) = ?$
8. Solve for m. $5m + 3 = 9m - 1$
9. Write the formula for finding the area of a trapezoid.
10. $(-9)^4 = ?$
11. How many tons are 18,000 pounds?
12. Simplify. $12xy^{-3}$ (Rewrite without using negative exponents.)
13. $43\frac{1}{10} + 26\frac{2}{5} = ?$
14. Solve for x. $\frac{9}{12} = \frac{x}{144}$
15. What value of x makes the sentence true? $-8x = 120$
16. 25% of 60 is what number?
17. Change from standard to scientific notation. 0.001002
18. $0.81 - 0.5673 = ?$
19. Solve for x. $\frac{1}{8}x + 14 = 20$
20. On Friday night, the waiters pooled their tips. The tips for the night totaled $175.65. Five waiters had worked the entire shift on Friday. How much did each waiter receive if the total was divided equally?

1.	2.	3.	4.
5.	6.	7.	8.
9.	10.	11.	12.
13.	14.	15.	16.
17.	18.	19.	20.

Lesson #47

1. Solve the inequality and graph the solution on a number line. $12b-5>-29$

2. Stan, the baker, reduced his weekly flour order from 8 lb. to 5 lb. By what percent has his order been reduced? (Round to the nearest whole number.)

3. Solve to find the value of h. $7h+2h-3=15$

4. Rewrite in standard notation. 1.7×10^{-13}

5. Two bowling balls are selected from a rack. The rack has 5 blue, 3 yellow, 6 green, and 2 purple bowling balls. What is the P(green and blue) with replacement? P(purple and green) without replacement?

6. Simplify. $x^{-5}y^{7}$

7. $102+(-67)=?$

8. Solve for x. $\dfrac{x}{14}=-8$

9. How many quarts are in 15 gallons?

10. Solve the system using any method. $\begin{array}{l}3x-y=4\\x+5y=-4\end{array}$

11. Solve for m. $16m+4=12+8m$

12. A line passing through points (–2, 4) and (3, 9) has what slope?

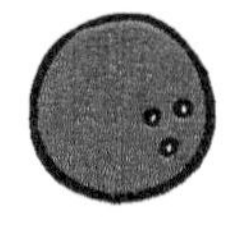

13. Evaluate $xy+3y$ when $x=3$ and $y=4$.

14. Which is greater, $\dfrac{4}{25}$ or 18%?

15. $1\dfrac{1}{2}\cdot\dfrac{4}{9}=?$

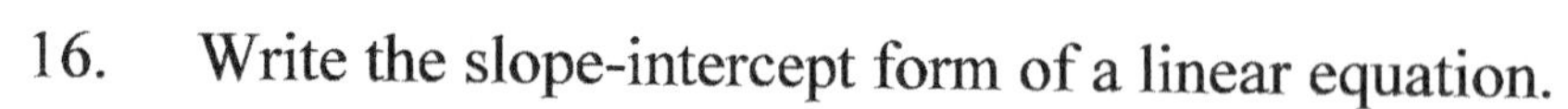

16. Write the slope-intercept form of a linear equation.

17. $0.0007\times0.006=?$

18. Translate into an algebraic phrase: *Nineteen divided by a number less three.*

19. $\sqrt{3364}=?$

20. Simplify. $\dfrac{18a^{2}bc^{4}}{24ab^{3}c^{2}}$

1.	2.	3.	4.
5.	6.	7.	8.
9.	10.	11.	12.
13.	14.	15.	16.
17.	18.	19.	20.

Lesson #48

1. Write 4^{-2} as a simple fraction.

2. Give the equation of the line through the point (4, 0) with a slope of 7.

3. Solve the system of equations using any method. $\begin{aligned} 6x+y&=13 \\ y-x&=-8 \end{aligned}$

4. Solve for x. $2(5x-1)=18$

5. Write 1.38×10^4 in standard notation.

6. What is the value of x in $x+21=-77$?

7. Find the volume of the rectangular prism.

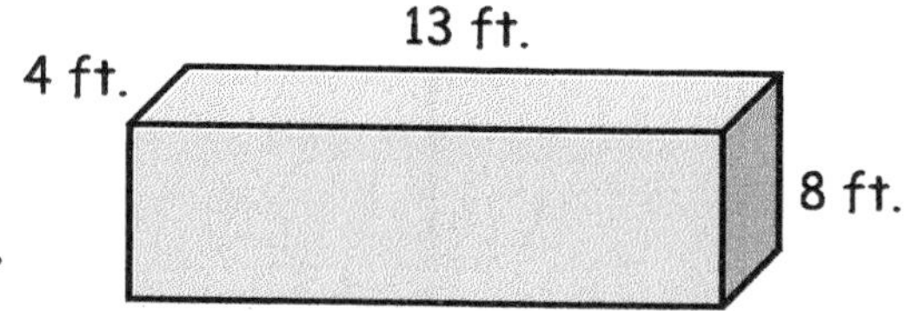

8. Solve for a. $17a+12=8a-24$

9. What values of b make the inequality true? $4-b>15$

10. $18+3[2(10\div2)-1]=?$

11. $-7(14)=?$

12. Jenny plans to buy either a cell phone, a CD player, or an mp3 player. All are available in either black or silver and can be stored in either her bedroom, her bathroom, or the family room. How many different options are there?

13. $(3.74)^0=?$

14. How many degrees are in a circle?

15. Solve for x. $\frac{x}{5}=16$

16. What is the circumference of a circle if the diameter is 14 mm?

17. Simplify. $6(2x+3y-7)-2(3x+4y+5)$

18. 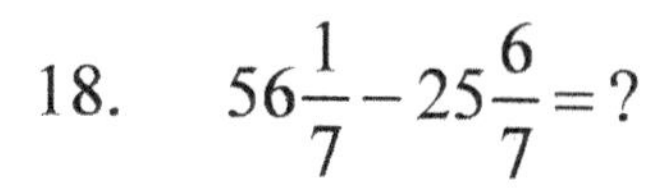 $56\frac{1}{7}-25\frac{6}{7}=?$

19. Water freezes at ________ °F.

20. Find the values of y in the equation $y=x-7$ when $x=\{0, -7, 5\}$.

1.	2.	3.	4.
5.	6.	7.	8.
9.	10.	11.	12.
13.	14.	15.	16.
17.	18.	19.	20.

Lesson #49

1. Find the solution set for this inequality. $6x - 15 \le 9$ or $10x > 84$

2. Simplify. $a^6b^3 \cdot a^2b^{-2}$

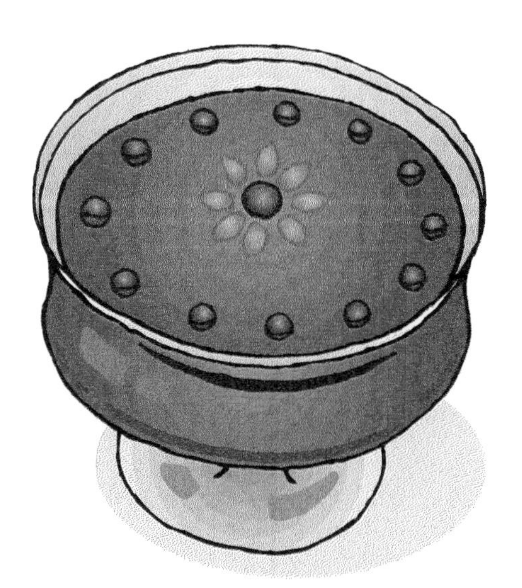

3. Evaluate $\frac{2x - y}{5}$ when $x = 4$ and $y = 3$.

4. Solve for x. $\frac{4}{15} = \frac{x}{45}$

5. Solve for b. $8b + b - 5 = 67$

6. $6 \cdot 5 + 40 \div 8 - 3 + 2 = ?$

7. Simplify. $\frac{1}{x^{-1}}$

8. Mrs. Jordan added $1\frac{3}{8}$ quarts of soda to the $2\frac{1}{2}$ quarts of fruit juice in the punch bowl. How many quarts of punch mixture did she make?

9. $\frac{-420}{-60} = ?$

10. Write 44,909 in scientific notation.

11. $\begin{pmatrix} 3 & -9 \\ -4 & 0 \end{pmatrix} + \begin{pmatrix} -4 & -6 \\ -2 & 5 \end{pmatrix} = ?$

12. Simplify. $(a^2b^3)(a^5)$

13. Use any method of your choice to solve the system. $y = 2x$; $7x - y = 35$

14. Write 0.56 as a reduced fraction and as a percent.

15. The slope of a horizontal line is __________.

16. Find the perimeter of the pentagon.

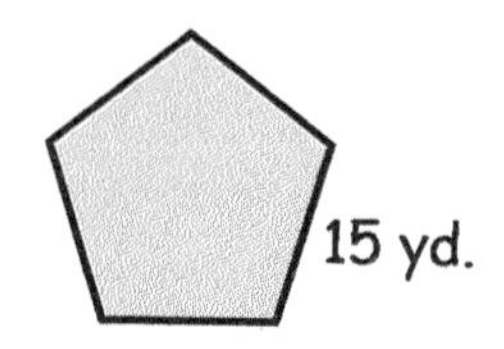

17. Write the equation of the line passing through points (–8, 2) and (1, 3).

18. Solve for x. $6x - 4 = 20$

19. How many cups are in 21 pints?

20. Put these decimals in decreasing order. 0.75 0.005 0.57 0.705

1.	2.	3.	4.
5.	6.	7.	8.
9.	10.	11.	12.
13.	14.	15.	16.
17.	18.	19.	20.

Lesson #50

1. Write 4,000,000,000,000 in scientific notation.

2. $2.5 \times 3.4 = ?$

3. Write the first 4 prime numbers.

4. Solve the inequality and graph the solution on a number line. $37 < 3c + 7 < 43$

5. A triangle with no congruent sides is called ___________.

6. $4 + 5(6) - (7 + 8) - 9 = ?$

7. Solve for c. $\frac{c}{9} = -12$

8. Ciera is making pancakes for breakfast. She has to mix $1\frac{3}{4}$ cups of whole wheat flour, $1\frac{1}{3}$ cups of white flour, and $\frac{1}{2}$ cup of buckwheat flour. How much flour will she use all together?

9. $119 - (-36) = ?$

10. Simplify. $x^6 \cdot y^2 \cdot x^4$

11. How many feet are in 6 miles?

12. What number is 60% of 50?

13. $-97 \bigcirc -21$

14. What is the formula for finding the area of a circle?

15. Simplify. $x^{-5}y^{-6}$

16. $75\frac{1}{3} + 86\frac{2}{5} = ?$

17. Solve for h. $15h - 7 = 8h$

18. Find the GCF of $15a^2b^2c^2$ and $35ab^2c^3$.

19. What value of b makes the sentence true? $b - 17 = 24$

20. Simplify. $b^5(b^3)^2$

1.	2.	3.	4.
5.	6.	7.	8.
9.	10.	11.	12.
13.	14.	15.	16.
17.	18.	19.	20.

Lesson #51

1. Write 0.00000078 in scientific notation.

2. Find $\frac{5}{6}$ of 72.

3. What is the value of $xy - \frac{x}{y}$ when $x = 4$ and $y = 2$?

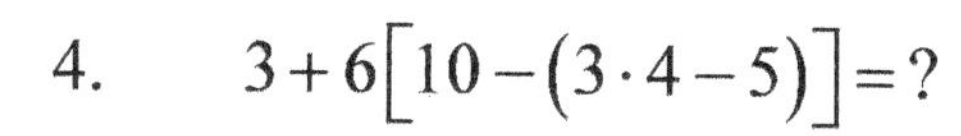

4. $3 + 6\left[10 - (3 \cdot 4 - 5)\right] = ?$

5. 48% of 25 is what number?

6. $\frac{5}{8} \times \frac{12}{15} = ?$

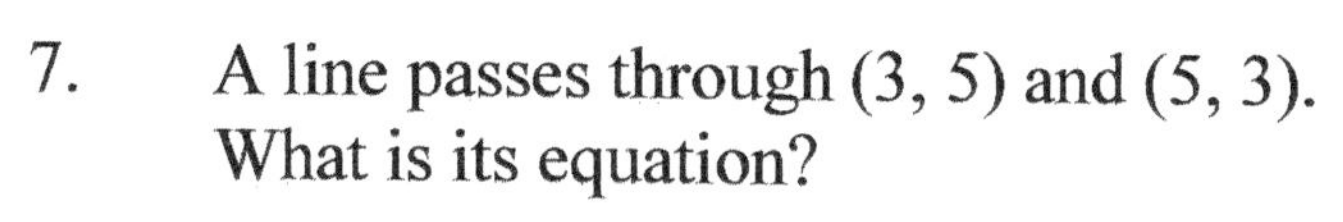

7. A line passes through (3, 5) and (5, 3). What is its equation?

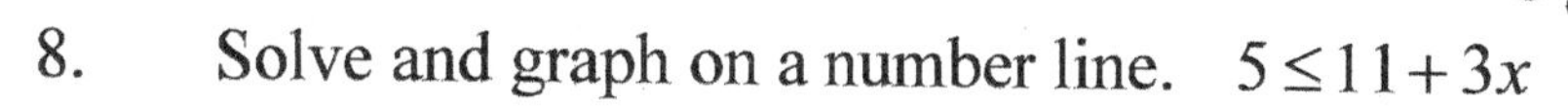

8. Solve and graph on a number line. $5 \leq 11 + 3x$

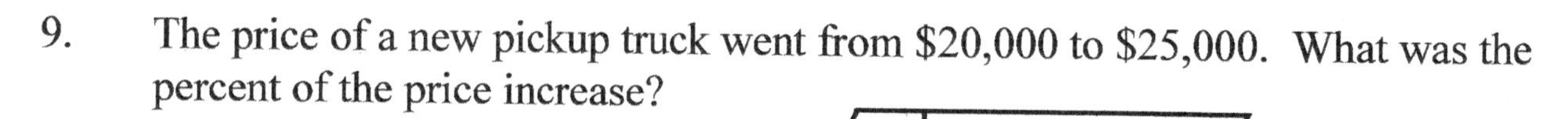

9. The price of a new pickup truck went from \$20,000 to \$25,000. What was the percent of the price increase?

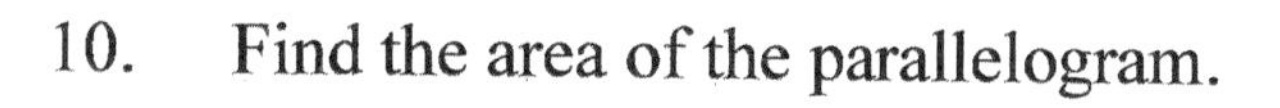

10. Find the area of the parallelogram.

30 cm

32 cm

11. Solve for c. $-7c = 105$

12. $1.84 \div 0.4 = ?$

13. $296.7 + 38.892 = ?$

14. What is the P(H, H, H, H, T, T, T) on 7 flips of a coin?

15. Round 47,816,275 to the nearest million.

16. Simplify. $(-5)^4 = ?$

17. How many ounces are in 8 pounds?

18. Simplify. $\frac{x^{-2}}{y^{-3}}$

19. $-33 - (-18) = ?$

20. What is the value of x? $\frac{x}{14} + 9 = 17$

1.	2.	3.	4.
5.	6.	7.	8.
9.	10.	11.	12.
13.	14.	15.	16.
17.	18.	19.	20.

Lesson #52

1. Evaluate c^{-a} when $a = 3$ and $c = -4$.

2. Write 1.6×10^6 in standard form.

3. $-73 + (-56) + 28 = ?$

4. A triangle with 2 congruent sides is a(n) ________ triangle.

5. What percent of 40 is 24?

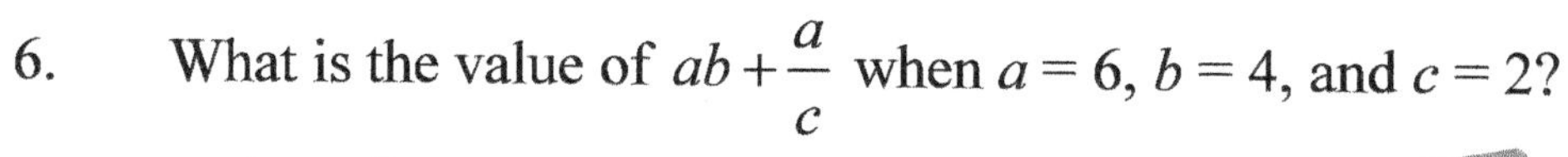

6. What is the value of $ab + \frac{a}{c}$ when $a = 6$, $b = 4$, and $c = 2$?

7. $14\frac{1}{2} - 9\frac{3}{4} = ?$

8. Solve for a. $6a + 30 = -36$

9. Simplify using only positive exponents. $(5x)^{-4}$

10. Solve for x. $\frac{12}{16} = \frac{9}{x}$

11. $9h > -108$ Graph the solution to the inequality.

12. Calculate the circumference of the circle.

12 cm

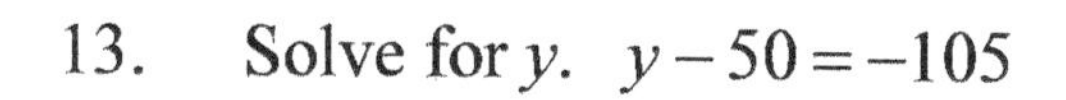

13. Solve for y. $y - 50 = -105$

14. Simplify. $(x^6)^3(y^2)$

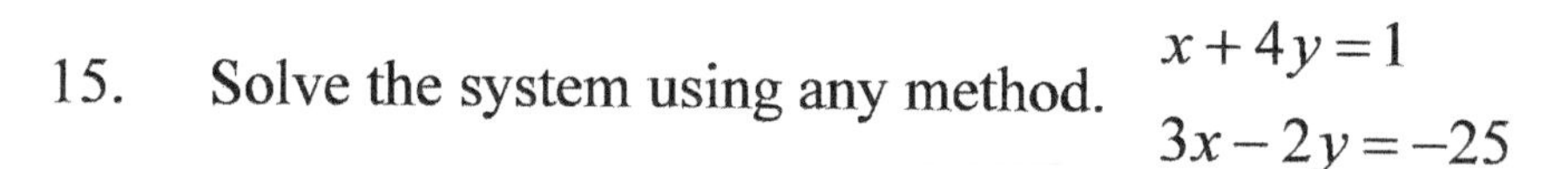

15. Solve the system using any method. $x + 4y = 1$; $3x - 2y = -25$

16. How many centimeters are in 8 meters?

17. What value of x makes the equation true? $3x + 4 = x + 18$

18. $7 + 7 \cdot 7 - 7 \div 7 = ?$

19. Kendra's mom chose six different candy bars to sell as a fundraiser for the school band. If there were 25 different candy bars in the catalog, what percentage of the possible choices did she select?

20. $5.216 \div 0.4 = ?$

1.	2.	3.	4.
5.	6.	7.	8.
9.	10.	11.	12.
13.	14.	15.	16.
17.	18.	19.	20.

Lesson #53

1. Simplify. $\dfrac{b^4}{b^9}$

2. Write 34,000,000 in scientific notation.

3. $-118-(-75)=?$

4. Solve for x. $\dfrac{x}{12}+8=12$

5. What is the value of a in $7a=80+9a$?

6. A line passes through point (4, 6) with a slope of –5. Write its equation.

7. Solve for w. $-8w<24$

8. What value of b makes the equation true? $3b-8=-32$

9. $\begin{pmatrix} 9 & -5 & -3 \\ 2 & 7 & 8 \end{pmatrix} - \begin{pmatrix} 7 & -3 & 2 \\ 0 & -5 & 3 \end{pmatrix} = ?$

10. Find the perimeter of a heptagon whose sides measure 4 ft. each.

11. How many pints are in 14 quarts?

12. Solve for c. $c+22=80$

13. –81 ◯ –77

14. $6.1-3.664=?$

15. Find the volume of this cylinder.

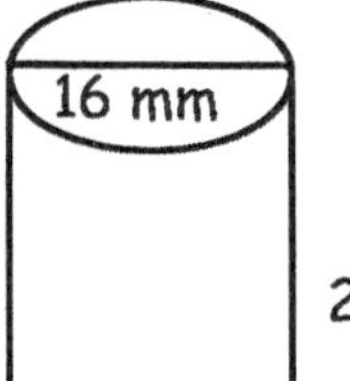

16. 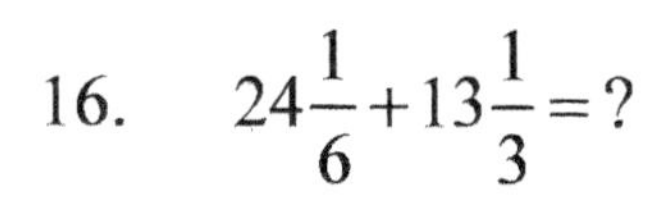 $24\dfrac{1}{6}+13\dfrac{1}{3}=?$

17. Use any method to solve. $\begin{aligned} x &= y-3 \\ x+2y &= 3 \end{aligned}$

18. Solve for c. $|c-4|=21$

19. Simplify. $c^{-2}c^{7}$

20. Give the coordinates for points A and C.

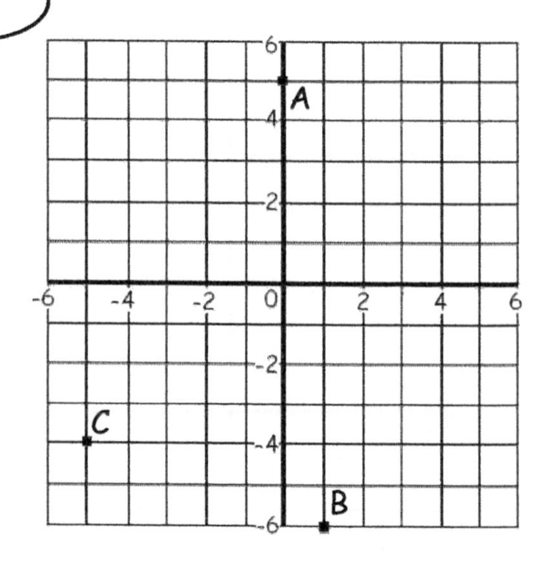

1.	2.	3.	4.
5.	6.	7.	8.
9.	10.	11.	12.
13.	14.	15.	16.
17.	18.	19.	20.

Lesson #54

1. Write 0.04×10^5 in correct scientific notation.

2. Simplify. $(x^5y^2)(x^{-6}y)$

3. $124 + (-58) = ?$

4. Simplify. $\frac{a^2b}{a^4b^3}$

5. $29 - 14\frac{4}{5} = ?$

6. Find the slope of the line through (0, 0) and (–7, 1).

7. Solve for x. $x + 5 = 3x - 7$

8. Simplify. $(a^2b^4)^3$

9. Find the LCM of $10x^3y^2$ and $15x^5yz^2$.

10. $6.165 \div 9 = ?$

11. Write 0.65 as a percent and as a reduced fraction.

12. Graph the solution to $8 \le 6 - m$ on a number line.

13. How many feet are in 5 miles?

14. Write 15×10^5 in proper scientific notation.

15. Mrs. Hughes put 25% of the fudge on the refreshment table at the beginning of the party. If there were 40 pieces of fudge on the table, how many pieces of fudge were there altogether?

16. What is the value of x in $7x - 3 = 18$?

17. Solve for y. $\frac{15}{9} = \frac{20}{y}$

18. Write an expression to represent *eighteen divided by a number increased by twice another number.*

19. Write in simplest form. $\frac{c^{12}}{c^{15}}$

20. Use the method of your choice to solve the system. $\begin{aligned} y &= x + 1 \\ y &= 2x - 1 \end{aligned}$

1.	2.	3.	4.
5.	6.	7.	8.
9.	10.	11.	12.
13.	14.	15.	16.
17.	18.	19.	20.

Lesson #55

1. Write 12,000,000,000 in scientific notation.

2. What number is $\frac{5}{6}$ of 72?

3. Solve the equation to find the value of m. $8m = 416$

4. On a trip from New York to Erie, Mr. Valentino drove an average of 60 miles each hour. If his trip took eight hours, how many miles did he drive?

5. $9^2 - [50 - (9 \cdot 5 - 30)] = ?$

6. What is the equation of the line through point (3, 0) having a slope of –1?

7. Solve for x. $\frac{3}{4}x = 24$

8. Simplify. -7^3

9. Evaluate $ab - (a - b)$ when $a = 5$ and $b = 2$.

10. What value of x makes the equation true? $\frac{1}{10}x + 19 = 25$

11. Solve for c. $c + 2.56 = 4.73$

12. Simplify. $(3n^4)^2$

13. A circle has a radius of 5 inches. What is its area?

14. Simplify. $a^7b^2 \cdot 21a^{-6}$

15. Solve the system using any method. $\begin{aligned} 2y &= x + 3 \\ x &= y \end{aligned}$

16. What are the values for y in the equation $y = 6x - 4$ when $x = \{2, 0, -2\}$?

17. $\sqrt{\frac{9}{16}} = ?$ (Hint: Find the square root of the numerator and the denominator.)

18. $400{,}000 - 216{,}877 = ?$

19. Simplify. $\frac{2^5}{2^7}$

20. Simplify. $\frac{5x^5}{15x^3}$

1.	2.	3.	4.
5.	6.	7.	8.
9.	10.	11.	12.
13.	14.	15.	16.
17.	18.	19.	20.

Lesson #56

1. $-132+(-116)=?$

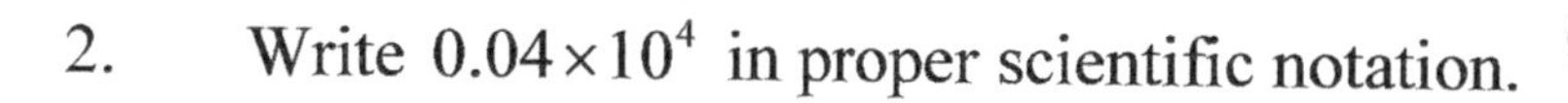

2. Write 0.04×10^4 in proper scientific notation.

3. $100-3[2(6-2)]=?$

4. $15\frac{1}{6}-10\frac{5}{6}=?$

5. Simplify. $\frac{c^{-1}d^3}{c^5d^{-4}}$

6. $0.009\times0.009=?$

7. $|x+4|\geq8$ Solve and graph the solution on a number line.

8. Give the slope-intercept form for a linear equation.

9. Solve for y. $4-y=10$

10. 64 is what percent of 80?

11. Simplify. $(4a^2b)^3(ab)^3$

12. What is the probability of rolling a two, a six, and a four on 3 rolls of a die?

13. What is the value of $ab+a+\frac{a}{b}$ when $a=6$ and $b=3$?

14. Find the range of the data in the stem-and-leaf plot.

15. Which is greater, 85% or $\frac{13}{20}$?

16. Simplify. $\frac{m^{-1}n^2}{m^3n}$

17. Solve for x. $\frac{x}{9}=-12$

8th Grade Bowling Scores

Stem	Leaf
6	3 7
7	0 3 8
8	5 9 9
9	6 6 6 9
10	0 8
12	1 7

18. Graph the solution to $-4<r-5\leq-1$ on a number line.

19. Write the formula for finding the surface area of a cube or a rectangular prism.

20. How many kilograms are the same as 6,000 grams?

1.	2.	3.	4.
5.	6.	7.	8.
9.	10.	11.	12.
13.	14.	15.	16.
17.	18.	19.	20.

Lesson #57

1. $10+10\cdot10-10\div10=?$

2. Simplify. $\dfrac{3x^{-2}}{y}$

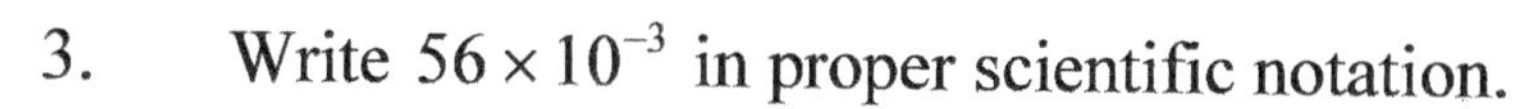

3. Write 56×10^{-3} in proper scientific notation.

4. $237+(-86)=?$

5. Simplify. $(7x^{5})(8x)$

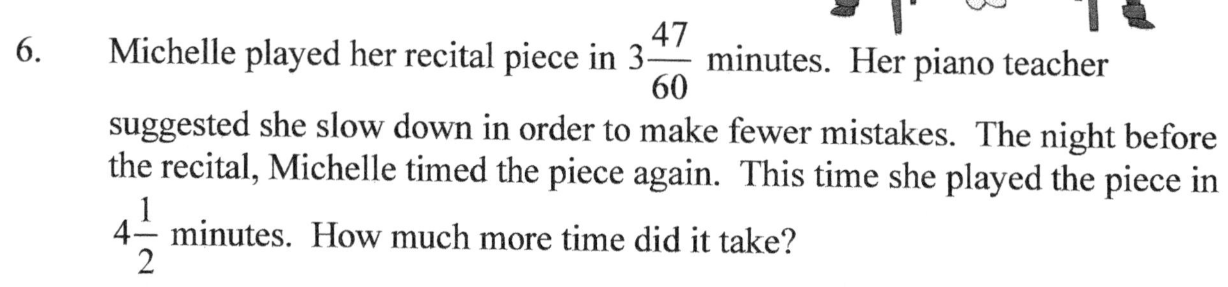

6. Michelle played her recital piece in $3\frac{47}{60}$ minutes. Her piano teacher suggested she slow down in order to make fewer mistakes. The night before the recital, Michelle timed the piece again. This time she played the piece in $4\frac{1}{2}$ minutes. How much more time did it take?

7. Solve for x. $6x-5=25$

8. $|y-2|\geq1$ Graph the solution.

9. How many ounces are in 9 pounds?

10. Simplify. $(3m^{3})^{4}$

11. $29\frac{1}{7}-13\frac{5}{7}=?$

12. If $a=4$, $b=3$, and $c=2$, what is the value of $ab+a-c$?

13. Calculate the area of the trapezoid.

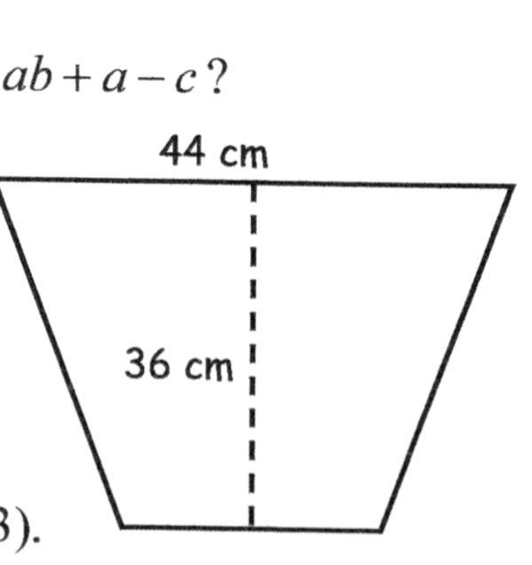

14. Solve for x. $9x+4=5x$

15. Simplify. $\dfrac{(2a^{7})(3a^{2})}{6a^{3}}$

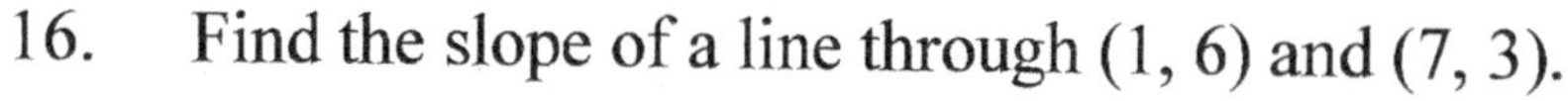

16. Find the slope of a line through (1, 6) and (7, 3).

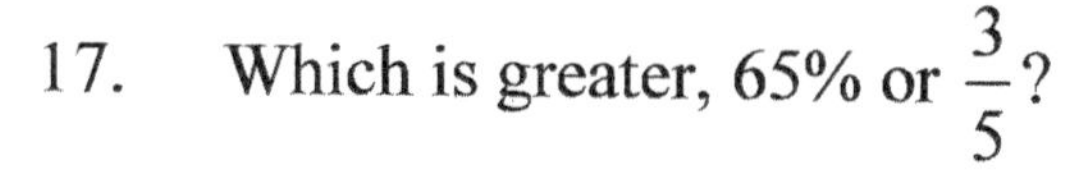

17. Which is greater, 65% or $\frac{3}{5}$?

18. Solve for x. $x+22=-60$

19. Solve the system $3x+y=3$ and $-3x+2y=-30$ using any method.

20. Simplify. $\left(\dfrac{4}{5}\right)^{3}$

1.	2.	3.	4.
5.	6.	7.	8.
9.	10.	11.	12.
13.	14.	15.	16.
17.	18.	19.	20.

Lesson #58

1. Rewrite in simplest terms. $\dfrac{6c^7}{3c}$

2. Solve for x. $\dfrac{x}{8}+6=12$

3. Write 65,000,000,000 in scientific notation.

4. Simplify. $2^{-6}=?$

5. What is $\dfrac{5}{9}$ of 36?

6. Simplify. $\dfrac{a^7b^3c^2}{a^2b^6c^2}$

7. $3^2+5\left[12-(8+1)\right]=?$

8. $\dfrac{5}{6}\bigcirc\dfrac{8}{9}$

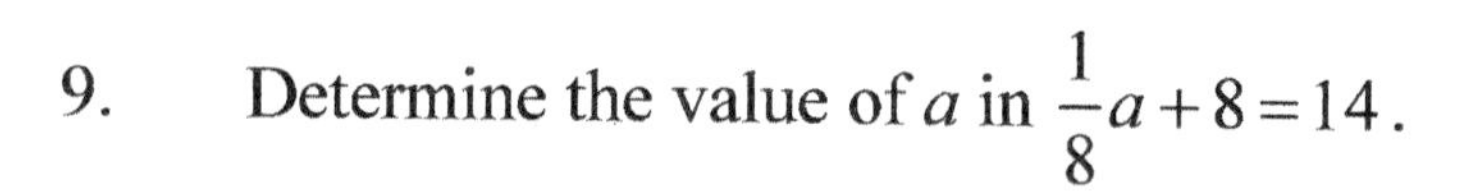

9. Determine the value of a in $\dfrac{1}{8}a+8=14$.

10. Solve the inequality for x. $-8x+16\le 8$

11. Write the slope-intercept form of a linear equation.

12. Simplify. $\left(g\cdot g^4\right)^2$

13. Calculate the area of the triangle.

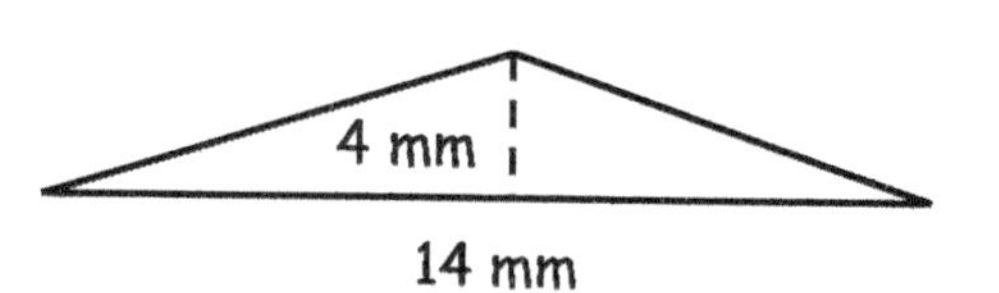

14. $-32+14+(-21)=?$

15. Solve for x. $\dfrac{2}{3}x=18$

16. Solve the system $2x-y=6$ and $-3x+4y=1$ using any method.

17. Solve for x. $5x=6x-19$

18. What value of x makes the equation true? $x+4.2=6.7$

19. $3.4\times 2.2=?$

20. Simplify. $\dfrac{1}{b^{-5}}$

1.	2.	3.	4.
5.	6.	7.	8.
9.	10.	11.	12.
13.	14.	15.	16.
17.	18.	19.	20.

Lesson #59

1. Simplify. $2a^2(3a+5)$

2. Write 42×10^6 in correct scientific notation.

3. $\frac{7}{8}\times\frac{16}{21}=?$

4. Simplify. $(5x^5)(3y^6)(3x^2)$

5. A line passes through points (7, 0) and (3, – 4). What is the slope?

6. $\begin{pmatrix}3 & -9\\0 & -2\end{pmatrix}-\begin{pmatrix}1 & -6\\3 & 2\end{pmatrix}=?$

7. Simplify. $(g^{10})^{-4}$

8. Solve the proportion for x. $\frac{5}{6}=\frac{x}{84}$

9. Solve for x. $-7x=84$

10. $0.4515\div0.5=?$

11. Find the LCM of $12x^3y^3z^4$ and $18x^2yz^2$.

12. $50-3[4(10-5)]=?$

13. Simplify. $(2^6)^0$

14. How many centimeters are in 7 meters?

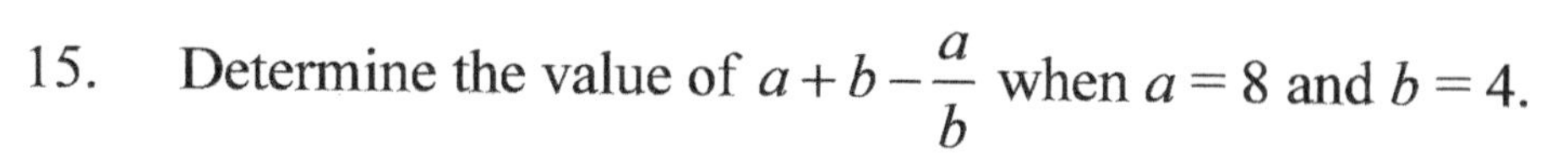

15. Determine the value of $a+b-\frac{a}{b}$ when $a=8$ and $b=4$.

16. Find the slope and the y-intercept for the line with equation. $y=6x+8$

17. Solve for x. $\frac{x}{12}=-5$

18. $72-46\frac{2}{3}=?$

19. Solve and graph the solution on a number line. $\frac{x}{4}>-1$

20. Two-fifths of the 30 children in the class have never played soccer. How many of them have not played before? How many have played before?

1.	2.	3.	4.
5.	6.	7.	8.
9.	10.	11.	12.
13.	14.	15.	16.
17.	18.	19.	20.

Lesson #60

1. Simplify. $(7cd^4)^2$

2. Mario's dog weighs 80 lbs. now. As a puppy, he weighed only 60 lbs. By what percent has the dog's weight changed?

3. Simplify. $\frac{2^7}{2^5}$

4. Solve for x. $x - 36 = 75$

5. Write 39×10^{-4} in correct scientific notation.

6. Simplify. $4(3a + 4b - 6) - 2(2a - 2b - 5)$

7. Find the value(s) of b. $4b > 36$

8. Evaluate $2x + 3y$ if $x = 4$ and $y = 2$.

9. Solve using any method. $y = -3x + 1$; $y = 3x + 7$

10. Write $\frac{7}{25}$ as a decimal and as a percent.

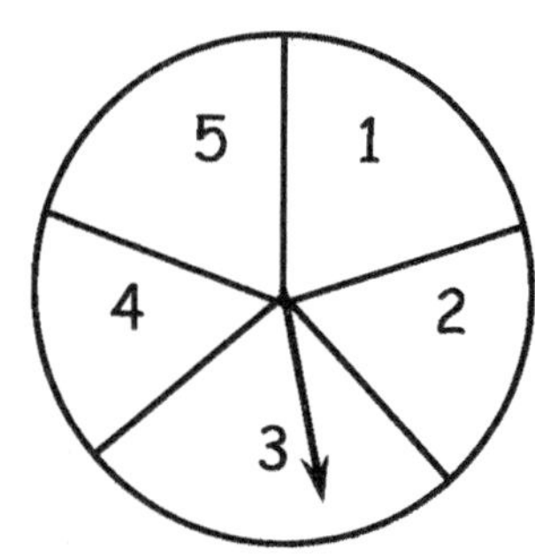

11. A student spins the spinner twice. On the first spin, the spinner lands on the number 5. On the second spin, the spinner lands on a prime number. The student says the probability for those two consecutive spins is $\frac{4}{5}$. Explain the error in this student's thinking and correct his answer.

12. Simplify. $\frac{a^{-21}a^{15}}{a^3}$

13. Give the equation of a line that includes point (–5, 2) and has a slope of 0.

14. Water boils at ________°C.

15. $900{,}000 - 799{,}816 = ?$

16. Solve for a. $3a - 7 = 20$

17. What value of x makes the equation true? $4x = -56$

18. $16 \div 2 + 5 \cdot 3 - 2 + 3^2 = ?$

19. How many years are 9 decades?

20. Solve to find the value of a. $8a - 3 = 2a + 15$

1.	2.	3.	4.
5.	6.	7.	8.
9.	10.	11.	12.
13.	14.	15.	16.
17.	18.	19.	20.

Lesson #61

1. Find the sum of the polynomials. $(3x^2 - 2x + 3) + (-2x^2 + 6x + 7)$

2. Write the formula for finding the volume of a cylinder.

3. Simplify. $(15a^3)(-3a)$

4. $38 - 16\frac{2}{5} = ?$

5. Solve for a. $10a - 4 = 6a$

6. Simplify. $c^4 \cdot d^3 \cdot c^2$

7. $(-148) + (-52) = ?$

8. Write 23×10^4 in correct scientific notation.

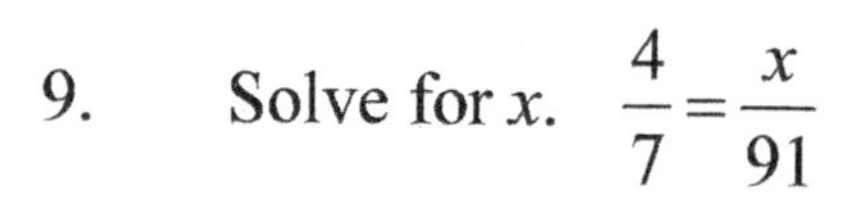

9. Solve for x. $\frac{4}{7} = \frac{x}{91}$

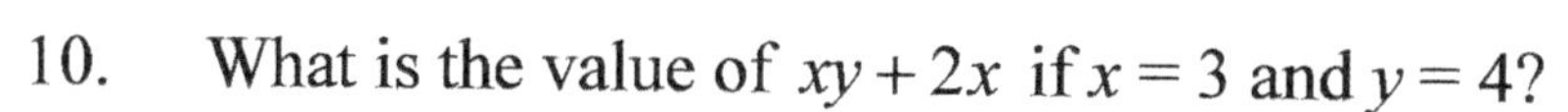

10. What is the value of $xy + 2x$ if $x = 3$ and $y = 4$?

11. Round 37.286 to the nearest tenth.

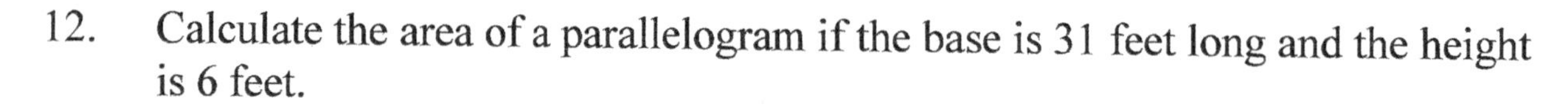

12. Calculate the area of a parallelogram if the base is 31 feet long and the height is 6 feet.

13. What number is 24% of 75?

14. $30 - 5[4 + 3 \cdot 2 - 5] = ?$

15. Write the formula for finding the circumference of a circle.

16. Simplify. $(2x^3)^2 = ?$

17. How many grams are 8 kilograms?

18. A triangle with three congruent sides is a(n) ______________ triangle.

19. Find the GCF of $12x^3y^2z$ and $18xy^2z^3$.

20. A bag contains 24 red marbles, 30 white marbles, and 40 blue marbles. What is the ratio of red marbles to blue marbles? What is the white to red ratio?

1.	2.	3.	4.
5.	6.	7.	8.
9.	10.	11.	12.
13.	14.	15.	16.
17.	18.	19.	20.

Lesson #62

1. Simplify. $\dfrac{a^7b^3c^2}{a^2b^6c^2}$

2. $-125(-5)=?$

3. Write 64,000,000,000 in scientific notation.

4. The slope of a horizontal line is ____________.

5. Simplify. $\left(\dfrac{3}{4}\right)^2$

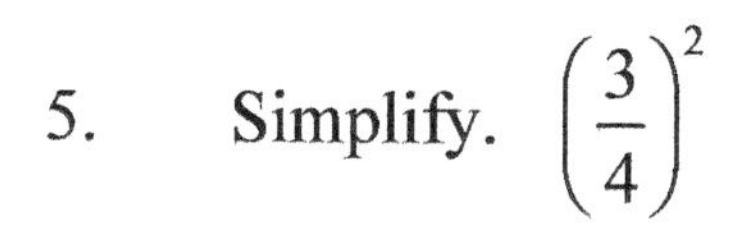

6. $9+9\cdot9-9\div9=?$

7. $\dfrac{5}{12}\cdot\dfrac{14}{25}=?$

8. What is the slope of a line through (1, 2) and (6, 1)?

9. When $a=2$, $b=3$, and $c=4$, evaluate $abc-2b$.

10. Simplify. $3x^2\cdot x^2=?$

11. Subtract the polynomials. $(7x^3-3x+1)-(x^3+4x^2-2)$

12. Find the perimeter of a decagon whose sides each measure 15 inches.

13. Solve the system using substitution. $\begin{aligned} x-y&=20 \\ 2x+3y&=0 \end{aligned}$

14. Write $\dfrac{6}{25}$ as a decimal and as a percent.

15. Solve for b. $b-124=316$

16. How many quarts are in 13 gallons?

17. If there are 60 students in Frank's class, how many more students like running than like football?

18. Determine the value of x in $\dfrac{x}{5}+15=30$.

19. Solve for m. $m+5(m-1)=13$

20. What is the probability of rolling a prime number with one roll of a die?

Frank's Class Survey

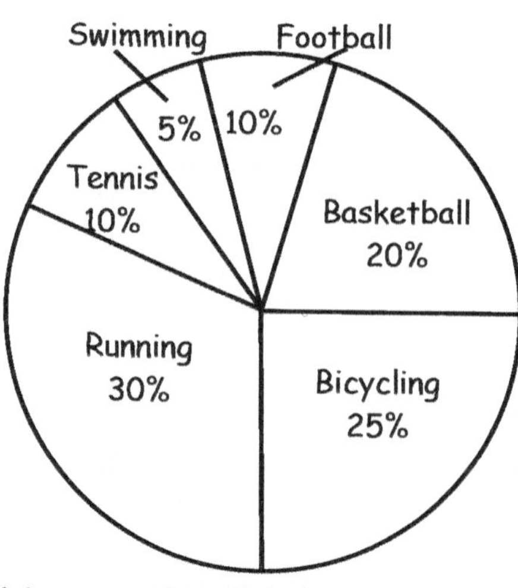

1.	2.	3.	4.
5.	6.	7.	8.
9.	10.	11.	12.
13.	14.	15.	16.
17.	18.	19.	20.

Lesson #63

1. Identify the slope and the y-intercept of the line $y = 7x - 4$.

2. $\sqrt{\frac{9}{16}} = ?$

3. What value of w makes the equation true? $\frac{w}{4} = 16$

4. Solve for x. $6x - 2 = x + 13$

5. What is the value of b in $\frac{4}{5}b = -20$?

6. $-46 - (-12) = ?$

7. Find the sum. $(7a^3 + 3a^2 - a + 2) + (8a^2 - 3a - 4)$

8. Simplify. $\frac{2^2}{2^5}$

9. Write 1.5×10^{-3} in standard form.

10. $86\frac{1}{4} - 27\frac{1}{2} = ?$

11. Simplify. $(5ac^3)^{-2}$

12. $5 \leq 11 + 3h$ Solve and graph on a number line.

13. Find the LCM of $8x^2y$ and $10x^3y^2z$.

14. $34.8 + 16.27 + 9.465 = ?$

15. $\begin{pmatrix} 7 & 2 & -3 \\ 5 & -1 & -4 \end{pmatrix} + \begin{pmatrix} 5 & -4 & -6 \\ 0 & -2 & 7 \end{pmatrix} = ?$

16. Bradley can swim 4 laps in 6 minutes. At this rate, how many minutes will he need to swim 10 laps?

17. $40 - 5[4 + 3(2 + 3) - 6] = ?$

18. Solve for n. $n - 22 = -66$

19. Which is greater, $\frac{2}{50}$ or 8%?

20. Write the equation of the line through point (3, 0) with a slope of -1.

1.	2.	3.	4.
5.	6.	7.	8.
9.	10.	11.	12.
13.	14.	15.	16.
17.	18.	19.	20.

Lesson #64

1. Solve for x. $\frac{6}{40}=\frac{15}{x}$

2. $\sqrt{100}-\sqrt{36}=?$

3. How many pints are in 8 quarts?

4. $\frac{-255}{-5}=?$

5. Find the difference. $\left(2x^3-5x^2+3x-1\right)-\left(8x^3-8x^2+4x+3\right)$

6. Find the area of the trapezoid to the right.

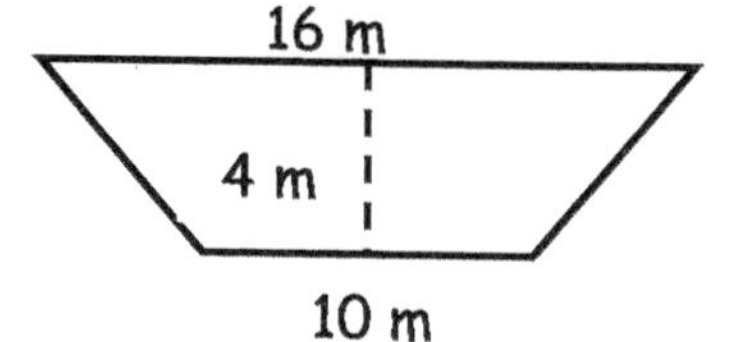

7. Evaluate $m+n+\frac{m}{n}$ when $m=8$ and $n=4$.

8. Which digit is in the ten-thousandths place in 26.0754?

9. $0.0008\times 0.004=?$

10. Write 1.6×10^7 in standard form.

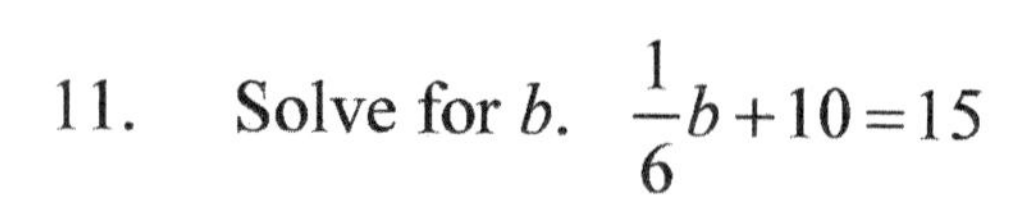

11. Solve for b. $\frac{1}{6}b+10=15$

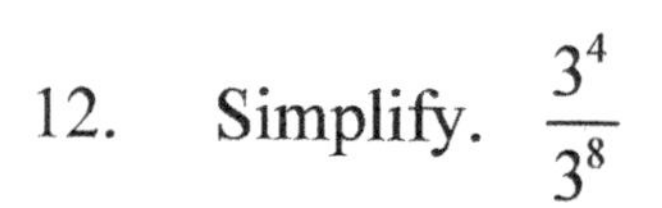

12. Simplify. $\frac{3^4}{3^8}$

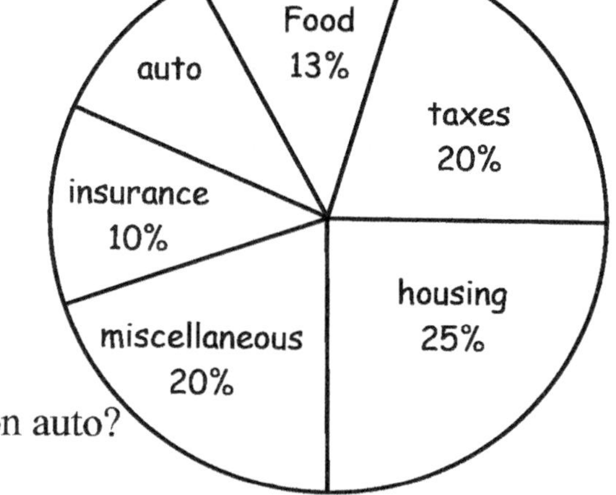

13. If the family's income was \$70,000, what amount of the family's income was spent on auto?

14. Simplify. $\left(5b^3\right)^4$

15. What geometric solid describes the shape of a soup can?

16. $4+a>3$ or $6a<-30$ Solve the inequality and graph on a number line.

17. $\sqrt{9}+\sqrt{16}-\sqrt{25}=?$

18. The slope of a vertical line is ________________.

19. Simplify. $\frac{c^3d^7}{c^8d^{-1}}$

20. $\frac{6}{10}\div\frac{2}{10}=?$

1.	2.	3.	4.
5.	6.	7.	8.
9.	10.	11.	12.
13.	14.	15.	16.
17.	18.	19.	20.

Lesson #65

1. $1.02 \div 0.12 = ?$

2. Simplify. $\dfrac{(5x^5)(2x^2)}{10x^2}$

3. Solve for a. $7a + 2 = 5a$

4. What is the P(H, T, H) on 3 flips of a coin?

5. Write 5.2×10^6 in standard form.

6. How many ounces are 7 pounds?

7. Give the slope-intercept form of a linear equation.

8. $\sqrt{\dfrac{25}{64}} = ?$

9. Simplify. $n^{10} \cdot n^{-4}$

10. What number is $\dfrac{5}{6}$ of 48?

11. Find the sum. $(7y^2 - 3y + 2) + (8y^2 + 3y - 4)$

12. $|-39| = ?$

13. Solve $5 + x > 9$ and graph its solution on a number line.

14. A nonagon has _______ sides.

15. Find the missing measurement, x.

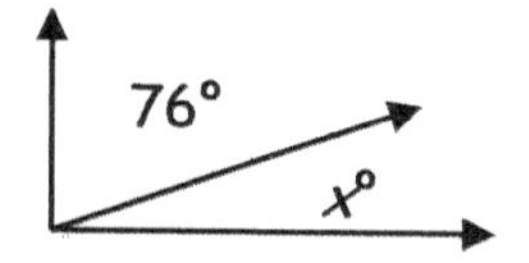

16. Write 75% as a decimal and as a reduced fraction.

17. Solve for x. $\dfrac{x}{7} = -12$

18. Write an expression for *the difference of a number and twenty-three.*

19. The stage director took a 22-ft. length of cord and divided it into $1\frac{3}{8}$-ft. lengths. How many cords were there?

20. An increase from 16 lbs. to 24 lbs. represents what percent of change?

1.	2.	3.	4.
5.	6.	7.	8.
9.	10.	11.	12.
13.	14.	15.	16.
17.	18.	19.	20.

Lesson #66

1. Simplify. $\left(\frac{5x^3}{20x}\right)^3$

2. $8+8\cdot8-8\div8=?$

3. Write the formula for calculating the volume of a rectangular solid.

4. Solve for x. $x-19=-41$

5. Write 27,000 in scientific notation.

6. Simplify. $2(4a+3b)+4(3a-2b)$

7. Simplify. $\frac{w^2}{w^5}$

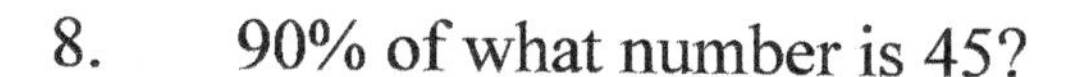

8. 90% of what number is 45?

9. During their last vacation, the King family traveled by train. They traveled 594 miles in 9 hours. What was the average speed of the train?

10. Find the difference. $(9a^3-3a^2+4a-2)-(6a^3-2a+6)$

11. $(-5)(-6)(-2)=?$

12. Write *a number divided by three, increased by fifteen* as an algebraic phrase.

13. $\begin{pmatrix}2 & -5\\-4 & 3\end{pmatrix}-\begin{pmatrix}1 & 7\\-3 & 2\end{pmatrix}=?$

14. Determine the area of the circle.

7 mm

15. Solve for a. $\frac{a}{9}+3=9$

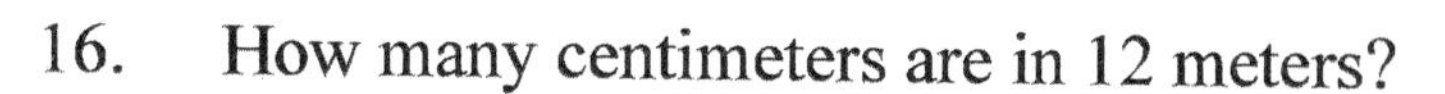

16. How many centimeters are in 12 meters?

17. Find the value of 2^8.

18. Simplify. $\frac{1}{2^{-3}}$

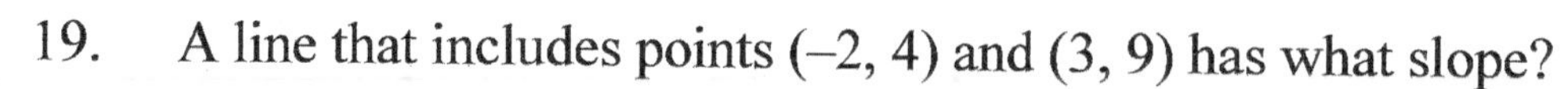

19. A line that includes points (–2, 4) and (3, 9) has what slope?

20. Solve for x. $7x+2<5x$

1.	2.	3.	4.
5.	6.	7.	8.
9.	10.	11.	12.
13.	14.	15.	16.
17.	18.	19.	20.

Lesson #67

1. Multiply. $x\left(5x^2+6x\right)$

2. Use your preferred method to solve the system. $\begin{array}{l} x-y=7 \\ 3x+2y=6 \end{array}$

3. $76+(-33)=?$

4. Solve the inequality and graph the solution. $\frac{x}{4}>-1$

5. Simplify. $\frac{a^2b^{-7}c^4}{a^5b^3c^{-2}}$

6. The price of a cheeseburger combo increased from \$4.50 to \$5.00. By what percent did the price change? (Round to the nearest whole number.)

7. Write 0.00063 in scientific notation.

8. Solve for x. $\frac{2}{5}=\frac{x}{65}$

9. Write 0.65 as a percent and as a reduced fraction.

10. $\sqrt{\frac{25}{4}}=?$

11. $\frac{7}{15}\cdot\frac{12}{21}=?$

12. Factor. $3x^3-9x^2+15x$

13. When $x=\{0,-1,3\}$, what are the values for y in $y=3x$?

14. Determine the sum of the polynomials. $\left(2x^2+3x-2\right)+\left(4x^2-5x+2\right)$

15. Simplify. $5m^5m^{-8}$

16. What is the value of a in the equation? $4a+8=24$

17. $15+3\left[4(8-3)-2\right]=?$

18. Evaluate $xy+x-y$ when $x=4$ and $y=3$.

19. Simplify. 4^{-3}

20. Solve for x. $\frac{1}{7}x-5=10$

1.	2.	3.	4.
5.	6.	7.	8.
9.	10.	11.	12.
13.	14.	15.	16.
17.	18.	19.	20.

Lesson #68

1. Simplify. $\left(x^2n^4\right)\left(n^{-8}\right)$

2. Factor. $3m^2+9m-6$

3. $74-38\frac{5}{8}=?$

4. Put these integers in increasing order. –46, –3, –12, –21

5. Write 1500 in scientific notation.

6. $-16+(-18)+7=?$

7. $0.6-0.341=?$

8. Solve for a. $a-12=60$

9. 80% of what number is 56?

10. Find the product. $4b\left(b^2+3\right)$

11. Solve the inequality for x. $x-6\leq-4$

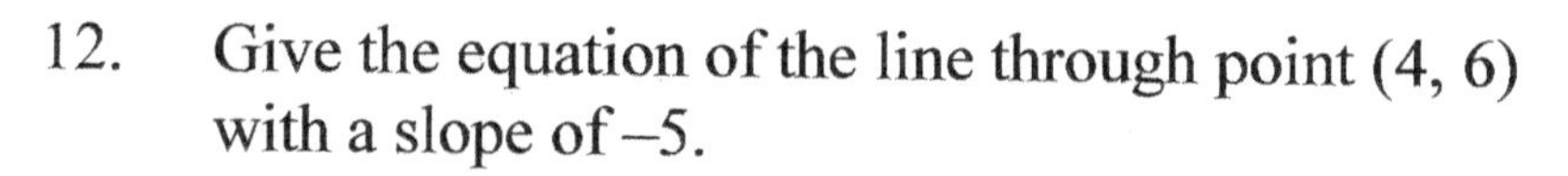

12. Give the equation of the line through point (4, 6) with a slope of –5.

13. When $b=2$ and $c=-4$, evaluate b^c.

14. Two angles whose measures add up to 180° are ________________ angles.

15. Find the difference. $\left(8x^3+6x^2-5\right)-\left(3x^3-2x^2+3\right)$

16. Solve for x. $15x+24=3x$

17. Simplify. $\dfrac{x^2yz^4}{xy^4z^{-3}}$

18. Lenny could do 75 push-ups in 3:45 (3 minutes and 45 seconds). How long does it take him to do one push-up? Give your answer in seconds.

19. What value of x makes the equation true? $\dfrac{x}{11}+3=9$

20. Determine the value of $a+b-ac$ if $a=6$, $b=5$, and $c=2$.

1.	2.	3.	4.
5.	6.	7.	8.
9.	10.	11.	12.
13.	14.	15.	16.
17.	18.	19.	20.

Lesson #69

1. Multiply using the FOIL method. $(2x+1)(x-5)$

2. $22\frac{1}{8}+36\frac{1}{2}=?$

3. Solve to find the value of a. $9a+12=6a-12$

4. 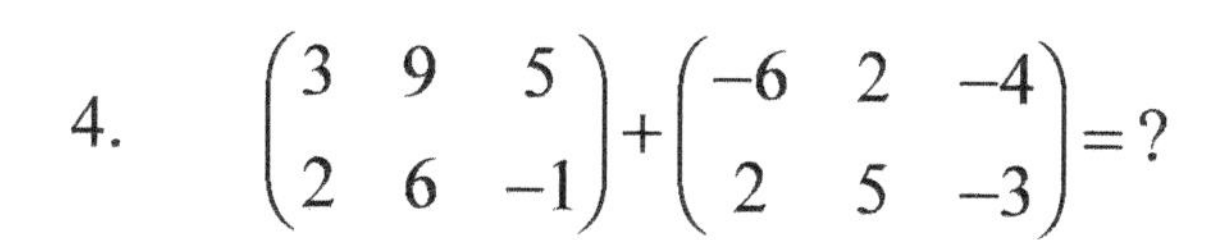
$\begin{pmatrix}3 & 9 & 5\\ 2 & 6 & -1\end{pmatrix}+\begin{pmatrix}-6 & 2 & -4\\ 2 & 5 & -3\end{pmatrix}=?$

5. Evaluate c^a when $a=3$ and $c=-4$.

6. For the linear equation $y=3x-9$ write the slope and the y-intercept.

7. Write $\frac{7}{20}$ as a decimal and as a percent.

8. Solve for x. $2x-9=9$

9. Simplify. $(3x^3)(2y)^2(3x^2)$

10. Simplify. 5^{-3}

11. Write 620,000 in scientific notation.

12. Water freezes at ________ °F.

13. Simplify. $\frac{m^{-2}}{m^{-5}}$

14. How many millimeters are in 17 meters?

15. Simplify. $(x^3y)^4$

16. Find the difference. $(7a^2+2a-5)-(2a^2-a-6)$

17. $700{,}000-216{,}893=?$

18. Solve for x. $\frac{x}{5}=-16$

19. Factor. $6a^2-8a$

20. Find the median, the mode and the range of 27, 39, 62, 12 and 27.

1.	2.	3.	4.
5.	6.	7.	8.
9.	10.	11.	12.
13.	14.	15.	16.
17.	18.	19.	20.

Lesson #70

1. Factor. $2x^2 - 4x^4$

2. Simplify. $(c^5)^2$

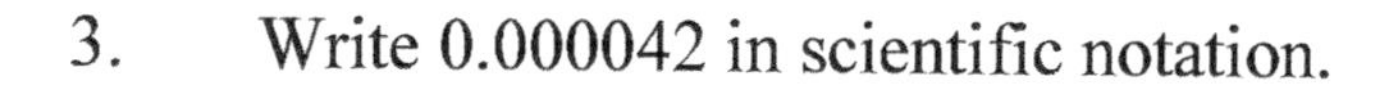

3. Write 0.000042 in scientific notation.

4. Solve for x. $-8x = 96$

5. Multiply using the FOIL method. $(3x-5)(2x+7)$

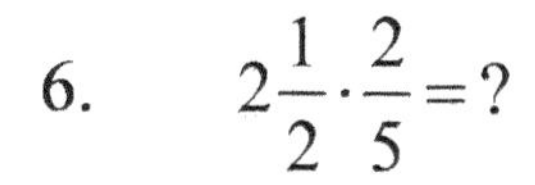

6. $2\frac{1}{2} \cdot \frac{2}{5} = ?$

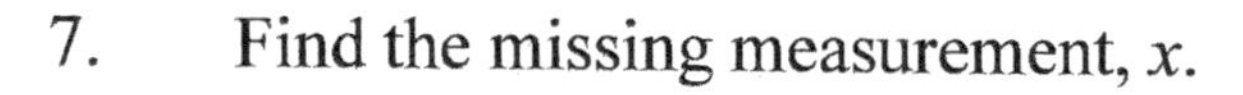

7. Find the missing measurement, x.

161° $x°$

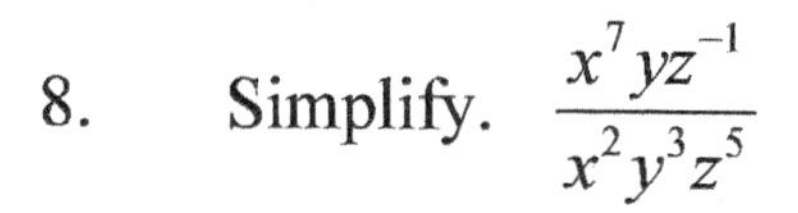

8. Simplify. $\dfrac{x^7 y z^{-1}}{x^2 y^3 z^5}$

9. $-|-79| = ?$

10. Simplify. $6x^2 \cdot 3y^3 \cdot 2y^4$

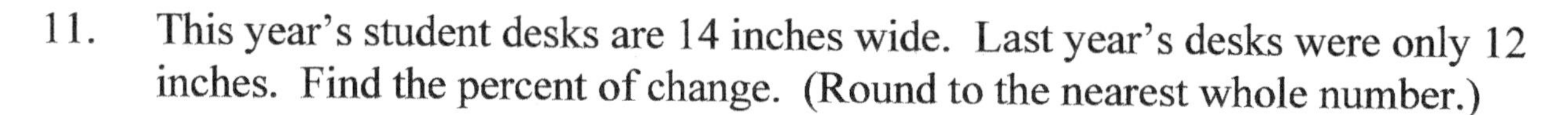

11. This year's student desks are 14 inches wide. Last year's desks were only 12 inches. Find the percent of change. (Round to the nearest whole number.)

12. Solve the inequality and graph the solution on a number line. $2x + 1 \leq -5$

13. The slope of a horizontal line is ____________.

14. $2.16 \times 0.05 = ?$

15. When $x = \{0, -1, 2\}$, find the values for y in the equation, $y = 6x - 4$.

16. $\sqrt{\dfrac{81}{144}} = ?$

17. $65 - 5 \cdot 9 + 18 - 12 \div 3 = ?$

18. Solve for b. $b - 13 = 27$

19. Evaluate $ab + ac$ when $a = 5$, $b = 2$, and $c = 3$.

20. Simplify. $(8x^3 - 2x^2 + 3x - 7) + (2x^3 + 2x^2 + 4x - 5) = ?$

1.	2.	3.	4.
5.	6.	7.	8.
9.	10.	11.	12.
13.	14.	15.	16.
17.	18.	19.	20.

Lesson #71

1. Solve for x. $14x-6=5x+12$
2. Factor. $9x+12x^2$
3. Multiply. $(3x^2+x-6)(2x-3)$
4. Simplify. $\frac{a^7}{a^9}$
5. Find the difference. $(9x^2-3x+2)-(5x+3)$
6. Find the volume of the cylinder.

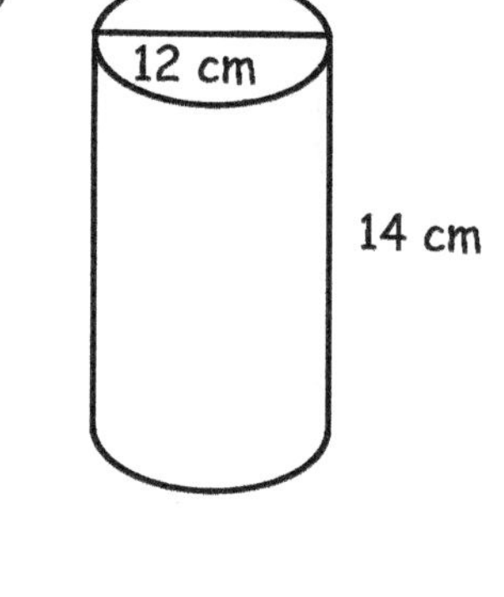

7. $14+2[3(5)-4+1]=?$
8. $\sqrt{196}=?$
9. Solve to find the value of x. $\frac{1}{9}x+4=12$
10. Use the FOIL method to multiply. $(x+3)(x+5)$
11. $39-21\frac{4}{5}=?$
12. $5(-7)(-2)=?$
13. Simplify. $4ab^0$
14. How many cups are in 13 pints?
15. Put these decimals in increasing order. 0.16 0.016 0.61 0.6
16. A line passing through points (4, 6) and (1, 8) has what slope?
17. $3a\geq 24$ Find the solution to the inequality and graph it.
18. Write 19,000,000 in scientific notation.
19. Find $\frac{2}{9}$ of 72.
20. Find the perimeter of a regular heptagon whose sides measure 12 inches.

1.	2.	3.	4.
5.	6.	7.	8.
9.	10.	11.	12.
13.	14.	15.	16.
17.	18.	19.	20.

Lesson #72

1. Factor. $5x^3 + 15x^2 + 10x$

2. Simplify. 5^{-3}

3. Simplify. $(a^2b^5)^2$

4. Which is greater, $\frac{4}{25}$ or 0.18?

5. Simplify. $(9a^3 + 2a^2 - 6) + (6a^3 - 2a^2 + 4a - 2)$

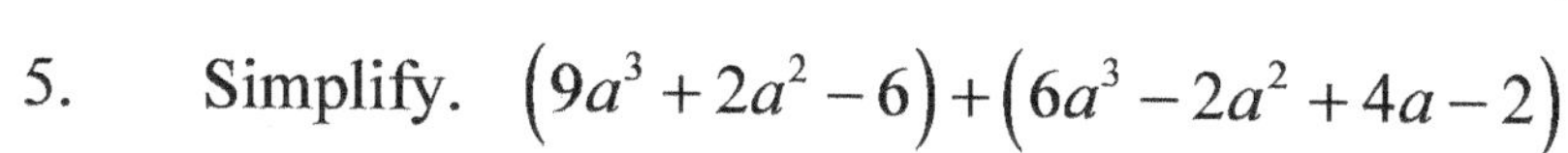

6. Find the product of $(5y - 8)$ and $(y + 4)$.

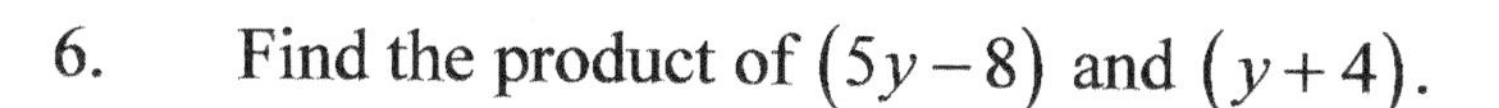

7. $-192 + (-188) = ?$

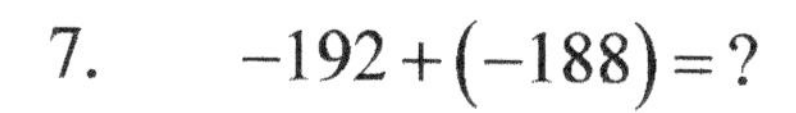

8. Simplify. $\frac{d^8}{d^5}$

9. Solve for x. $4x - 11 = 7x + 10$

10. Write the slope-intercept form of a linear equation.

11. A nonagon has ___________ sides.

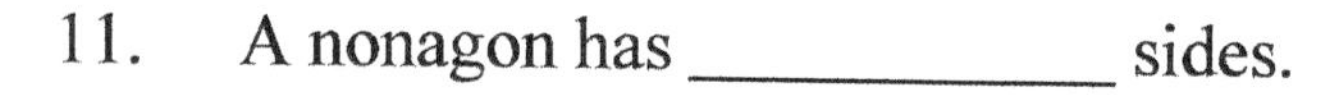

12. $42.86 + 9.3 = ?$

13. Write 0.0000078 in scientific notation.

14. What is the value of $x + 3y - y$ when $x = 6$ and $y = 2$?

15. Two angles whose measures add up to 90° are _________ angles.

16. Solve for b. $\frac{5}{6}b = -30$

17. Mr. Kendall planted 12 maple trees in his backyard. Only 75% of the trees survived the winter. How many trees survived?

18. Simplify. $(6x^3)(4x)$

19. Find the area of the trapezoid.

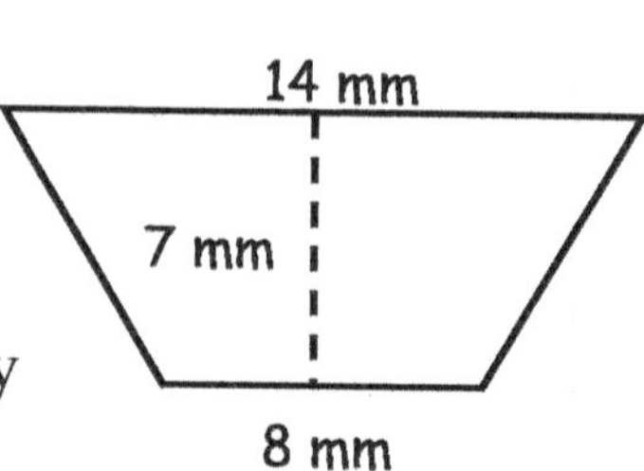

20. What is the probability of getting only heads on 5 flips of a coin?

1.	2.	3.	4.
5.	6.	7.	8.
9.	10.	11.	12.
13.	14.	15.	16.
17.	18.	19.	20.

Lesson #73

1. Simplify. $\dfrac{1}{a^{-4}}$

2. Factor. $6x^3 + 24x^2 + 12x$

3. Two angles whose measures add up to 180° are ____________ angles.

4. Evaluate $3a + 2b$ when $a = 4$ and $b = 3$.

5. $46\frac{1}{8} + 23\frac{2}{5} = ?$

6. Simplify. $\left(x^2y^3\right)^2$

7. Round 37.283 to the nearest tenth.

8. What is the value of x in $6x \geq 42$?

9. Find the difference.
$$\begin{array}{r} 8a^2 + 6a - 5 \\ - \;\; 4a^2 - 3a + 9 \\ \hline \end{array}$$

10. Write 730,000 in scientific notation.

11. Solve for x. $7x + 4 = 32$

12. Determine the slope of a line that passes through points (8, 3) and (9, 5).

13. Simplify. $\dfrac{a^{-3}}{a^{-7}}$

14. Find the product of $9c$ and $c^2 - 3c + 5$.

15. Write $\dfrac{2}{50}$ as a decimal and as a percent.

16. How many grams are in 4 kilograms?

17. What value of x makes the equation true? $\dfrac{x}{9} = -13$

18. 60% of the students in Mia's class are boys. There are 15 boys in the class. How many students are in the class?

19. Each bag of sour candy contains 16 pieces. Of these, three pieces are lime flavored. If Jordan has 144 pieces of lime candy, how many bags did he buy?

20. When $x = \{0, -4, 2\}$, what are the corresponding values of y in $y = 3x + 2$?

1.	2.	3.	4.
5.	6.	7.	8.
9.	10.	11.	12.
13.	14.	15.	16.
17.	18.	19.	20.

Lesson #74

1. Solve for p. $8p-3=13$

2. Determine the value of x. $\frac{4}{5}x=-20$

3. Solve the system using any method. $\begin{matrix} 3x-y=4 \\ x+5y=-4 \end{matrix}$

4. Find the slope and the y-intercept of the line whose equation is $y=-6x+5$.

5. $18+3[2+4\cdot 2]=?$

6. Find $\frac{5}{7}$ of 56.

7. The average weight of four polar bears is 624 pounds. The weight of the first polar bear is 626 pounds, the second bear weighs 645 pounds, the third bear weighs 610 pounds. What is the weight of the fourth polar bear?

8. A circle with a diameter of 12 centimeters has what area?

9. Solve for a. $\frac{1}{8}a+4=12$

10. $3.24\times 0.03=?$

11. Multiply these binomials. $(x+4)(x+7)$

12. Simplify. $(3x^4)^2$

13. Write in standard notation. 8.05×10^6

14. $70{,}000-31{,}966=?$

15. Factor. x^2+5x+6

16. Rewrite in simplest form. $\frac{5}{a^{-3}}$

17. Factor. $9c^2-169$

18. What value of b will make the equation true? $b+12=-50$

19. Find the GCF of $18a^2b^3c$ and $24ab^3c^2$.

20. Find the sum. $\begin{array}{r} 7x^3+3x^2-4x-6 \\ +\ 2x^3-2x^2-3x+2 \\ \hline \end{array}$

1.	2.	3.	4.
5.	6.	7.	8.
9.	10.	11.	12.
13.	14.	15.	16.
17.	18.	19.	20.

Lesson #75

1. Factor. $d^2 - 144$

2. $86.17 + 9.3 = ?$

3. Factor and solve for b. $b^2 + 3b - 4 = 0$

4. What is the percent of change from 18 ounces to 12 ounces?

5. What is the value of a if $\frac{1}{7}a + 14 = 20$?

6. **The standard form of a quadratic equation is written: $ax^2 + bx + c = 0$.** Write the standard form of a quadratic equation.

7. Solve the quadratic equation to find the value(s) of x. $2x^2 - 98 = 0$

8. Solve for x. $4x + 8 = 9x + 18$

9. Simplify. $\frac{w^7}{w^{-6}}$

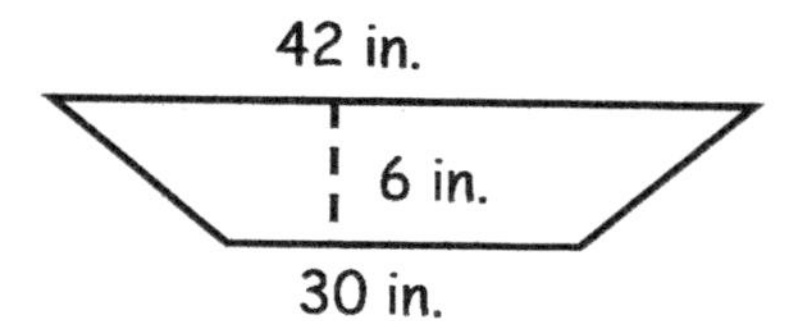

10. Determine the area of the trapezoid.

11. When $a = -5$ and $b = 8$, what is the value of $\frac{ab}{4}$?

12. Use any method to solve the system. $\begin{aligned} 4x - 5y &= 9 \\ -2x - y &= -29 \end{aligned}$

13. $81^0 = ?$

14. Solve and graph the solution. $t - 5 \leq -3$

15. Find the median and the mode of 83, 62, 54, 91 and 62.

16. The cost for parking in a parking deck can be figured using the following formula: $cost = 5 + 3(time - 2)$. What is the cost of parking in the deck for 6 hours? (Hint: Cost is measured in dollars and time, in hours.)

17. Multiply. $(x + 5)(x - 3)$

18. Write 1.8×10^5 in standard form.

19. How many years are in 5 centuries?

20. Find the difference. $(6a^3 - 3a^2 + 4a - 5) - (3a^3 - 2a + 3)$

1.	2.	3.	4.
5.	6.	7.	8.
9.	10.	11.	12.
13.	14.	15.	16.
17.	18.	19.	20.

Lesson #76

1. $43 - 25\frac{4}{9} = ?$

2. Factor the polynomial. $x^2 + 6x + 9$

3. Which digit is in the ten thousandths place in 62.4079?

4. A line passes through points (2, 5) and (4, 8). What is the slope of the line?

5. Factor. $d^2 - 144$

6. Simplify. t^{-3}

7. Write 3.2×10^{-5} in standard notation.

8. Find the value of 3^5.

9. $\frac{9}{10} \div \frac{3}{5} = ?$

10. Multiply. $(x+6)(x-2)$

11. Simplify. $\frac{(3x^4)(2x^2)}{6x^2}$

12. Solve for a. $8a + 6 = 14a$

13. Solve the system of equations. $\begin{matrix} 3x + 2y = 12 \\ x - 2y = -4 \end{matrix}$

14. Write 0.45 as a percent and as a reduced fraction.

15. A decagon has ___________ sides.

16. Find the area of the parallelogram.

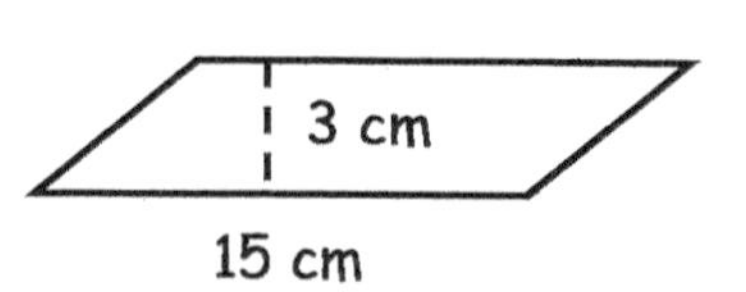

17. $5 + 5 \cdot 5 - 5 \div 5 = ?$

18. Translate *four times a number divided by three times another number* into an algebraic expression.

19. Write the formula for finding the surface area of a cube or rectangular prism.

20. $\begin{pmatrix} 4 & 7 & 3 \\ -2 & 0 & -5 \end{pmatrix} + \begin{pmatrix} 3 & -6 & -2 \\ 4 & -5 & -3 \end{pmatrix} = ?$

1.	2.	3.	4.
5.	6.	7.	8.
9.	10.	11.	12.
13.	14.	15.	16.
17.	18.	19.	20.

Lesson #77

1. Factor. $x^2 - 25$

2. Solve for c. $c + 25 = -70$

3. Simplify. 5^{-3}

4. How many quarts are in 7 gallons?

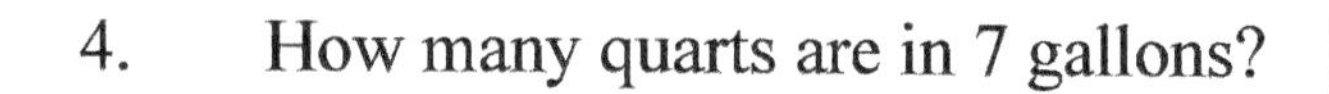

5. Simplify. $\left(x^2y\right)^2$

6. $50 - 25 \div 5 + 3 \cdot 4 - 1 = ?$

7. Write 0.00035 in scientific notation.

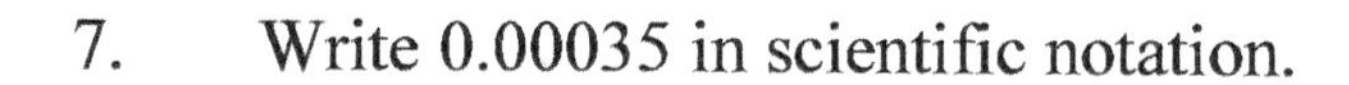

8. Find the GCF of $15a^2b^3c$ and $25abc$.

9. Write the equation for the line through point (3, –1), having a slope of 1.

10. Which is greater, $\frac{4}{5}$ or 0.65?

11. Factor. $7x^2 - 14x^3$

12. What is the probability of rolling a number greater than 2 on one roll of a die?

13. Simplify. $\frac{\left(4a^6\right)\left(3a^2\right)}{6a^3}$

14. Multiply $\left(2x^2 + x - 5\right)$ and $\left(2x - 4\right)$.

15. Solve the system using the method of your choice. $\begin{array}{l} x + y = 19 \\ x - y = -7 \end{array}$

16. $0.4 - 0.113 = ?$

17. A seven-sided shape is a(n) _______________.

18. Find the difference. $\begin{array}{r} 9a^3 - 6a^2 + 2a - 4 \\ -\ \ 5a^3 - 2a^2 + 4a - 3 \\ \hline \end{array}$

19. $\begin{pmatrix} -5 & 0 \\ 3 & -2 \end{pmatrix} - \begin{pmatrix} -2 & -6 \\ 3 & 4 \end{pmatrix} = ?$

20. **The quadratic formula is** $x = \frac{-b \pm \sqrt{b^2 - 4ac}}{2a}$. Write the quadratic formula.

1.	2.	3.	4.
5.	6.	7.	8.
9.	10.	11.	12.
13.	14.	15.	16.
17.	18.	19.	20.

Lesson #78

1. $-88+(-56)=?$

2. How many yards are in 2 miles?

3. Latrice is putting water into her aquarium using a pitcher which holds 2 quarts of fluid. Approximately how many pitchers-full will it take to fill the aquarium with at least 15 gallons of water?

4. Solve for x. $9 \le 7-x$

5. Simplify. $\frac{x^2y}{x^4y^2}$

6. Factor. $x^2-9x+20$

7. Solve for x. $\frac{4}{5}=\frac{x}{105}$

8. $8 \cdot 4 + 12 \div 3 - 8 - 1 = ?$

9. Which is greater, $\frac{3}{20}$ or 25%?

10. What are the values for y in the equation, $y=-7x$, when $x=\{-3, 0, 2\}$?

11. Write the formula for finding the area of a triangle.

12. Simplify. $(6x^4)(5x)$

13. If a line includes points (3, –6) and (5, 4), what is its slope?

14. $8\frac{1}{3}+3\frac{1}{4}=?$

15. Factor. c^2-100

16. Solve for x. $\frac{1}{5}x-3=12$

17. Simplify. -3^{-3}

18. Factor. $-3c^3+15c^2-3c$

19. Multiply. $(2y+1)(3y+4)$

20. Find the sum. $\begin{array}{r} x^2+x-3 \\ \underline{2x^2+4x-1} \end{array}$

1.	2.	3.	4.
5.	6.	7.	8.
9.	10.	11.	12.
13.	14.	15.	16.
17.	18.	19.	20.

Lesson #79

1. Factor. x^2+6x+8

2. $-124-(-88)=?$

3. Solve for x. $\frac{3}{7}x=21$

4. Simplify. $(n \cdot n^3)^2$

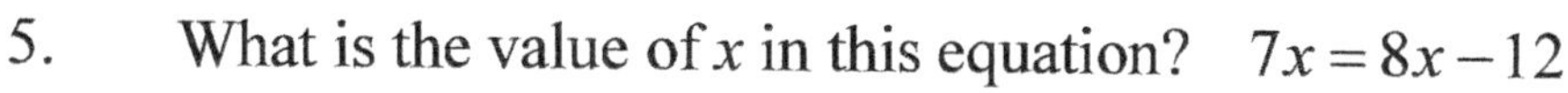

5. What is the value of x in this equation? $7x=8x-12$

6. Put these decimals in increasing order.
   2.74 2.47 2.07 2.7

7. Factor. y^2-64

8. Solve for a. $\frac{a}{7}=-12$

9. How many feet are in 4 yards?

10. $3.545 \div 0.05 = ?$

11. On four flips of a coin, what is the probability of getting tails, tails, tails and heads?

12. Find the volume of the cylinder.

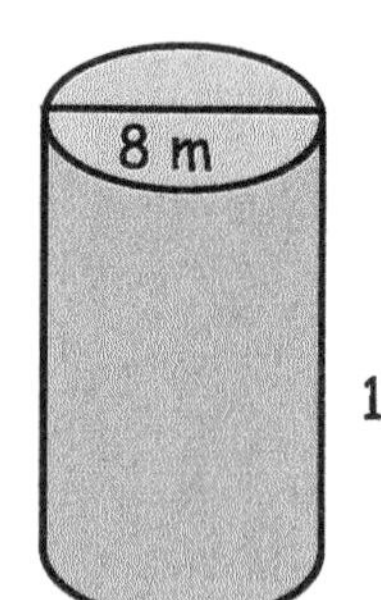

13. Multiply. $(x+7)(x-6)$

14. A nonagon has __________ sides.

15. Factor. a^2-5a+6

16. Write 42,000,000 in scientific notation.

17. Simplify. $c^{-4}d^2$

18. 90% of 60 is what number?

19. How many degrees are in a circle?

20. Find the difference.
$$\begin{array}{r} 3a^3+2a^2-5 \\ -\quad a^3+\ a^2+2 \\ \hline \end{array}$$

1.	2.	3.	4.
5.	6.	7.	8.
9.	10.	11.	12.
13.	14.	15.	16.
17.	18.	19.	20.

Lesson #80

1. Write $\frac{4}{25}$ as a decimal and as a percent.

2. When Gina bought a $330 leather jacket, the sales tax was 7.5%. What was the total cost of the leather jacket, including tax?

3. $42 \div 6 + 3[4 + 2 \cdot 2] = ?$

4. A triangle with two congruent sides is _____.

5. Solve for x. $7x - 5 = 16$

6. Write the slope and the y-intercept in the equation. $y = \frac{2}{3}x - 7$

7. Factor. $d^2 - 7d + 12$

8. $\frac{8}{9} \times \frac{18}{24} = ?$

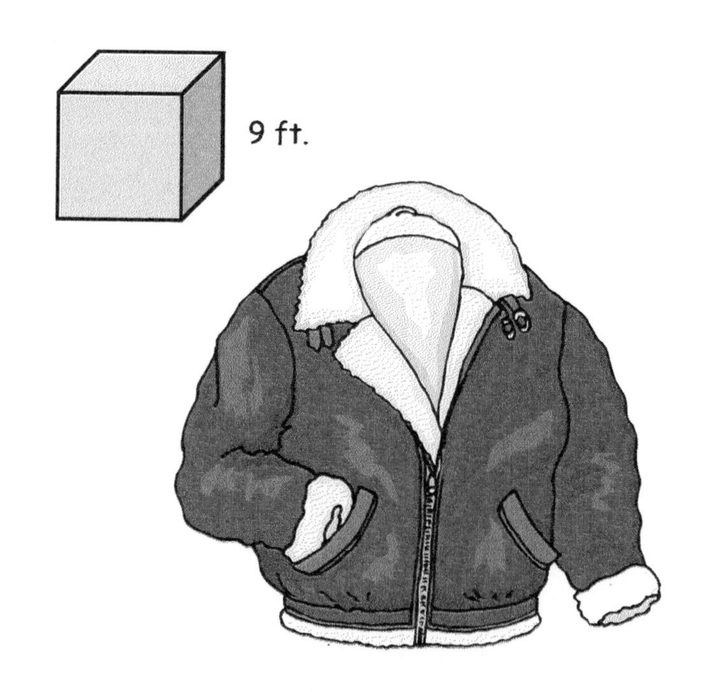

9. Find the volume of the cube.

10. Multiply. $(a-8)(a-9)$

11. Simplify. $x^{-4}y^{-3}$

12. $-93 \bigcirc -78$

13. $(-62) + (-34) + (15) = ?$

14. Factor. $10a^3 + 5a^2 + 5a$

15. Simplify. $4ac^0$

16. $\frac{x}{3} > -2$ Graph the solution on a number line.

17. Use the quadratic formula to solve $x^2 + 5x + 6 = 0$. (Hint: See Lesson #77.)

18. Simplify. $\frac{(3x^2)(5x^3)}{5x}$

19. Solve the system for x and y. $\begin{aligned} x + y &= 13 \\ -x + y &= -8 \end{aligned}$

20. Find the difference: $\begin{array}{r} 10a^3 + 2a^2 - 7 \\ -\ \ 6a^3 - \ \ a^2 - 4 \\ \hline \end{array}$

1.	2.	3.	4.
5.	6.	7.	8.
9.	10.	11.	12.
13.	14.	15.	16.
17.	18.	19.	20.

Lesson #81

1. Factor the GCF out of the trinomial. $6a^3 + 3a^2 - 18a$

2. $35\frac{1}{8} + 16\frac{2}{5} = ?$

3. Simplify. $8a^{-4}b^2c^{-3}$

4. Over his first 6 football games, Tyrone averaged 10 yards per carry. Over the next 9 games, he averaged 15 yards per carry. How many runs per carry did he average during all 15 of the games he played?

5. Solve the inequality for x. $8 \leq 12 - x$

6. Factor. $x^2 + 3x - 4$

7. Simplify. $\frac{8a^3b^2c^4}{12ab^2c}$

8. $(-35) + (-18) + (14) = ?$

9. Write 0.0000625 in scientific notation.

10. Simplify. $(3a^2b^3)^2$

11. Multiply. $(2x + 1)(7x - 4)$

12. $60 - 4[2 + 3 \cdot 4] = ?$

13. Use the quadratic formula (Lesson #77) to solve. $x^2 - 7x + 10 = 0$

14. Solve for x. $\frac{x}{18} = -7$

15. If a line includes points (7, 0) and (3, –4), what is its slope?

16. Write this equation in standard form. $3x^2 - 10 = -13x$

17. Find the area of the trapezoid shown to the right.

15 cm

12 cm

3 cm

18. What number is 25% of 60?

19. Solve the system of equations. $\begin{aligned} 3x + 7y &= 3 \\ x - 7y &= 1 \end{aligned}$

20. Find the difference. $\begin{array}{r} 6x - 4 \\ -\ \ 2x + 7 \\ \hline \end{array}$

1.	2.	3.	4.
5.	6.	7.	8.
9.	10.	11.	12.
13.	14.	15.	16.
17.	18.	19.	20.

Lesson #82

1. $56-33\frac{4}{9}=?$

2. $\begin{pmatrix} 6 & 0 & -3 \\ 4 & -7 & 5 \end{pmatrix}-\begin{pmatrix} -5 & -3 & 2 \\ 8 & 2 & -4 \end{pmatrix}=?$

3. $\frac{7}{9}\times\frac{18}{21}=?$

4. Simplify. $\left(m^7t^{-5}\right)^2$

5. Write 34,000,000 in scientific notation.

6. Multiply. $(a+5)(a+7)$

7. Solve for x. $6x-5=2x-21$

8. Simplify. $2y^4\cdot y^3$

9. Write the formula for finding the area of a triangle.

10. What is the value of a in the equation, $\frac{3}{4}a=15$?

11. 20% of what number is 15?

12. Put these integers in decreasing order. $-63\quad -3\quad -47\quad -12\quad -50$

13. What value of x will make the equation true? $4x-7=5$

14. Solve the inequality for x. $-3<2x+5\leq 9$

15. Evaluate $\frac{xy}{3}+2x$ when $x=5$ and $y=6$.

16. Ten out of every fifty fans cheer for the Rams. What percent of the fans cheer for the Rams? What fraction of the fans do not cheer for the Rams?

17. $(-6)(-2)(3)=?$

18. Write the slope-intercept form of a linear equation.

19. Solve for y. $|y-3|=7$

20. Find the sum. $\left(4y^2+6y+1\right)+\left(2y^2+3y-2\right)$

1.	2.	3.	4.
5.	6.	7.	8.
9.	10.	11.	12.
13.	14.	15.	16.
17.	18.	19.	20.

Lesson #83

1. Simplify. $4a^2(3a^5)$

2. Solve for x. $\frac{1}{7}x+15=25$

3. Use the quadratic formula to solve the equation, $c^2+6c+8=0$.

4. Write 0.084 in scientific notation.

5. What value of x makes these two fractions equivalent? $\frac{5}{9}=\frac{x}{117}$

6. Multiply. $(2x+5)(x+1)$

7. $2x-7>-7x+20$ Graph the solution on a number line.

8. Factor the GCF out of the trinomial. $8x^3-4x^2+12x$

9. Susan's salary was raised from $7 per hour to $9 per hour. Find the percent of increase in her hourly salary. Round to the nearest whole number.

10. Write the equation of the line through point (4, 0) with a slope of 7.

11. Find the median and the range of 62, 94, 75, 36 and 51.

12. Solve for x. $5-x=x+9$

13. When $x=\{0,-4,3\}$ in the equation $5x-8=y$, find the values for y.

14. What is the P(H, H, H, T, T, H) on 6 flips of a coin?

15. $42 \div 6+5 \cdot 2+8 \div 4=?$

16. Simplify. $\frac{7s^{-5}}{5t^{-3}}$

17. How many feet are in 4 miles?

18. Factor. $9w^2-16$

19. Solve the system. $\begin{aligned} 4x-y&=105 \\ x+7y&=-10 \end{aligned}$

20. Find the difference. $\begin{array}{r} 9y^3+4y-3 \\ -\ \ 8y^3-2y+5 \\ \hline \end{array}$

1.	2.	3.	4.
5.	6.	7.	8.
9.	10.	11.	12.
13.	14.	15.	16.
17.	18.	19.	20.

Lesson #84

1. Find the area of the circle with a radius of 14 millimeters.

2. $86.7 + 9.65 = ?$

3. Factor the trinomial. $a^2 + 12a + 36$

4. Simplify. $8x^{-4}y^3z^{-2}$

5. What are the slope and the y-intercept in the equation, $y = \frac{-2}{5}x + 6$?

6. Write 2.4×10^4 in standard notation.

7. Multiply. $(x+4)(x-4)$

8. A vending machine requires exact change. Julie wants to purchase some candy for 45¢. List all of the different combinations of dimes and nickels she could use to purchase the candy. (Hint: There are 5 combinations.)

9. $6\frac{1}{3} + 5\frac{2}{7} = ?$

10. Use the quadratic formula to solve. $x^2 - 9x + 14 = 0$

11. Solve for x. $x + 25 = -66$

12. Solve for y. $3y < 18$

13. Evaluate $x^{-3}y^2$ if $x = 2$ and $y = 4$.

14. Solve for y. $|2y - 4| = 14$

15. $-47 - (-19) = ?$

16. Calculate the perimeter of the rectangle.

23 in.

13 in.

17. Factor out the GCF of the three terms. $4x^2 - 12x - 36$

18. Simplify. $\frac{(3x^2)^3(2x)}{6x^2}$

19. Use the method of your choice to solve.

$$\begin{aligned} -3x + 4y &= 29 \\ 3x + 2y &= -17 \end{aligned}$$

20. Find the sum.

$$\begin{array}{r} 10x^3 - 6x^2 + 3x - 5 \\ +\ 4x^3 \qquad\quad - x + 3 \\ \hline \end{array}$$

1.	2.	3.	4.
5.	6.	7.	8.
9.	10.	11.	12.
13.	14.	15.	16.
17.	18.	19.	20.

Lesson #85

1. Simplify. 63^0

2. Factor. $st^2 - st - 20s$

3. Multiply. $(3x+4)(4x+5)$

4. Simplify. $(2b)(-6b^7)$

5. Put these decimals in increasing order. 3.42 3.2 3.042 3.24

6. Write 0.0000725 in scientific notation.

7. $18\frac{2}{5} - 12\frac{4}{5} = ?$

8. Use the quadratic formula to solve. $y^2 - 3y - 28 = 0$

9. Find $\frac{4}{7}$ of 28.

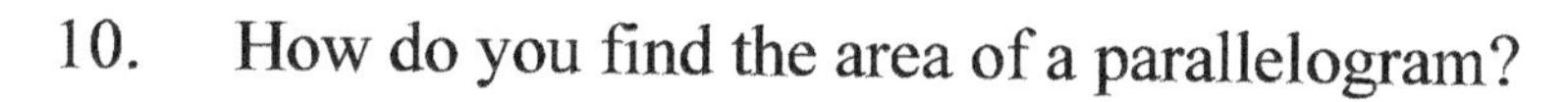

10. How do you find the area of a parallelogram?

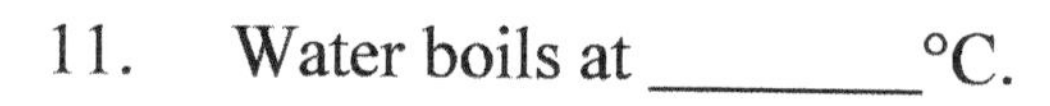

11. Water boils at ________ °C.

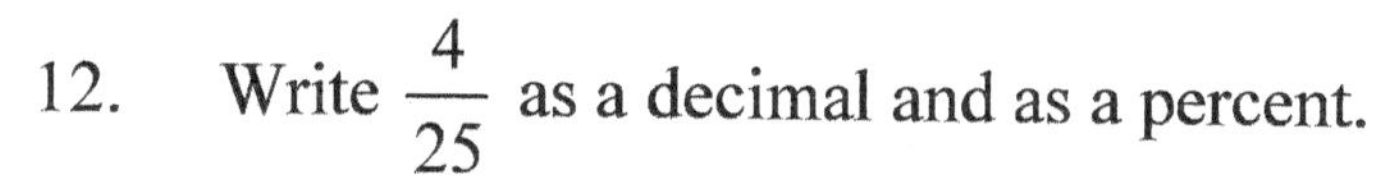

12. Write $\frac{4}{25}$ as a decimal and as a percent.

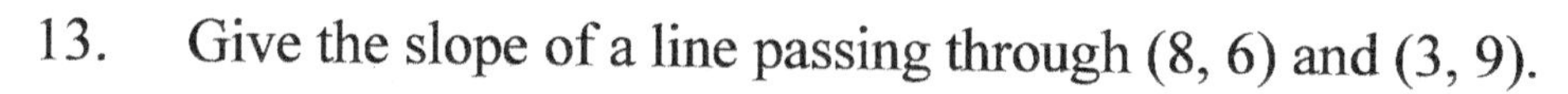

13. Give the slope of a line passing through (8, 6) and (3, 9).

14. Factor. $a^2 - 9a + 14$

15. $h + 4 \leq 12$ Solve the inequality and graph its solution.

16. On his first two tests Joel's average was 80 percent. On the next three tests he averaged 90 percent. What was Joel's average score for all 5 tests?

17. Simplify. $\frac{6x^2y^{-7}}{z^{-6}}$

18. What value of x makes the equation true? $\frac{4}{5}x = -20$

19. How many centimeters are in 60 meters?

20. Find the difference.

$$\begin{array}{r} 12x^3 + 8x^2 - 4x + 6 \\ -\quad 8x^3 - 3x^2 + 2x - 4 \\ \hline \end{array}$$

1.	2.	3.	4.
5.	6.	7.	8.
9.	10.	11.	12.
13.	14.	15.	16.
17.	18.	19.	20.

Lesson #86

1. Write the equation of a line passing through point (2, 3) with a slope of 3.

2. Write 0.00000734 in scientific notation.

3. Factor the trinomial. $b^2 - 4b - 21$

4. $66 - (-24) = ?$

5. $\begin{pmatrix} 3 & -4 \\ 8 & 0 \end{pmatrix} + \begin{pmatrix} -5 & -6 \\ -7 & 3 \end{pmatrix} = ?$

6. Solve for x. $4x = -64$

7. Write the quadratic formula.

8. $32 \div 4 + 6 \cdot 3 + 8 - 2 = ?$

9. Simplify. $7a^{-2}b^2c^{-4}$

10. Multiply. $5y(3y - 2)$

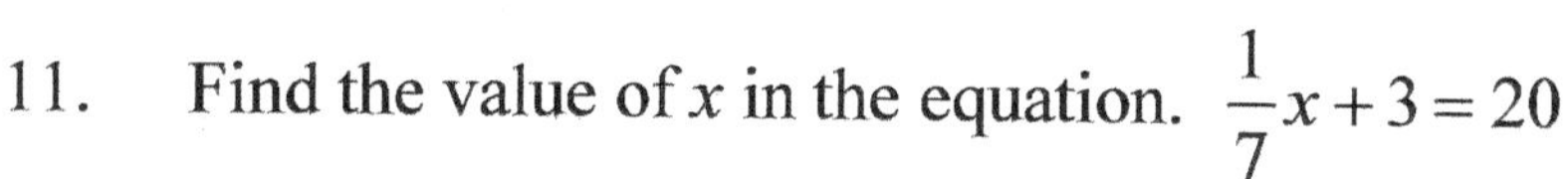

11. Find the value of x in the equation. $\frac{1}{7}x + 3 = 20$

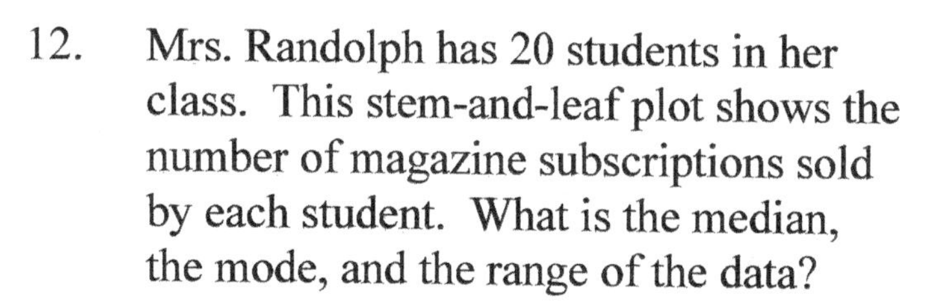

12. Mrs. Randolph has 20 students in her class. This stem-and-leaf plot shows the number of magazine subscriptions sold by each student. What is the median, the mode, and the range of the data?

Magazines Sold

Stem	Leaf
0	1 7 9
1	0 1 1 1 1 2 3 3 5 8 9
2	5 7
3	2 7 7
4	1

13. Simplify. $(3x^2y^3)^2$

14. Solve using the quadratic formula. $x^2 - 9x + 8 = 0$

15. $(-5)(-7)(2) = ?$

16. $16\frac{1}{7} - 8\frac{4}{7} = ?$

12 in.

17. Find the circumference of the circle.

18. Solve the inequality for x. $2x + 15 \le 21$

19. Write 0.75 as a reduced fraction and as a percent.

20. Find the difference.

$$\begin{array}{r} 5x^3 + 4y - 3 \\ -\ \ 8x^3 - 2y + 5 \\ \hline \end{array}$$

1.	2.	3.	4.
5.	6.	7.	8.
9.	10.	11.	12.
13.	14.	15.	16.
17.	18.	19.	20.

Lesson #87

1. What is the value of c in the equation? $7c-9=8c$

2. The slope of a horizontal line is always ____________.

3. Two angles whose measures add up to 180° are ____________ angles.

4. When $x=3$ and $y=4$, what is the value of $xy+2x$?

5. Multiply. $(c-6)^2$

6. 80% of what number is 72?

7. Dante's monthly electric bill went from \$60 to \$70. By what percent did his bill increase? Round to the nearest whole number.

8. Solve the inequality for a. $-15\le -3a+12$

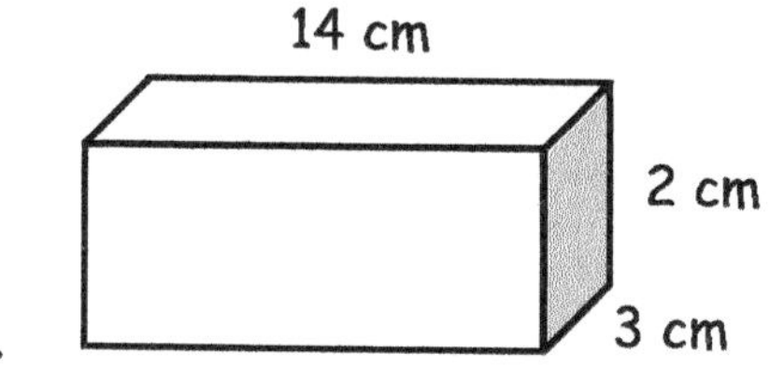

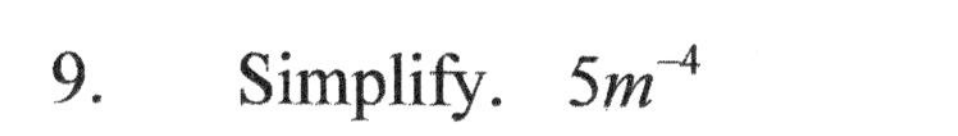

9. Simplify. $5m^{-4}$

10. Calculate the surface area of the rectangular prism.

11. Write the equation in standard form. $5x^2+5=12x$

12. Kiki bought a jacket for \$85. If the sales tax was 8%, what was the total price of the jacket?

13. Give the slope and the y-intercept for the equation. $y=\frac{2}{5}x-6$

14. Use the quadratic formula to solve. $x^2-7x+10=0$

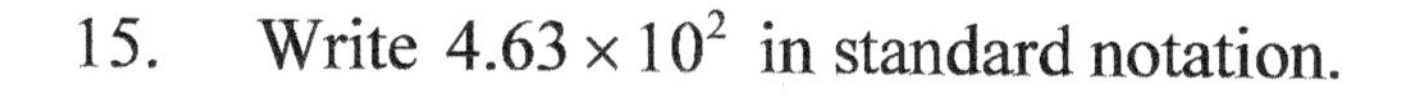

15. Write 4.63×10^2 in standard notation.

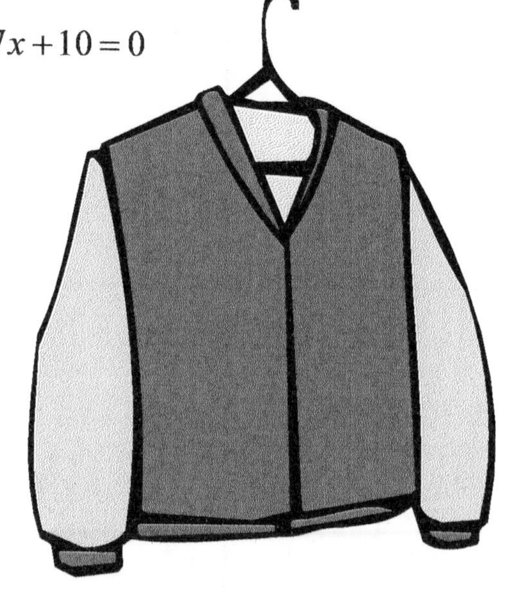

16. Factor out the GCF. $8a^3-6a^2+4a$

17. Solve for a. $\frac{3}{4}a=12$

18. What number is $\frac{3}{5}$ of 60?

19. Simplify. $\frac{16xy^2z^3}{18x^2y^2z^2}$

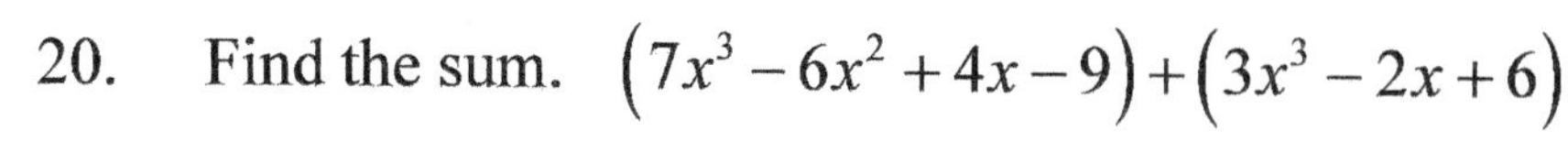

20. Find the sum. $\left(7x^3-6x^2+4x-9\right)+\left(3x^3-2x+6\right)$

1.	2.	3.	4.
5.	6.	7.	8.
9.	10.	11.	12.
13.	14.	15.	16.
17.	18.	19.	20.

Lesson #88

1. $40 - 3(5 \cdot 2) + 5 = ?$

2. Simplify. $5x^2y^{-4}$

3. Multiply. $4c\left(2c^2 - c + 3\right)$

4. What value of x makes the equation true? $5x - 1 = -26$

5. Write 16,000,000,000 in scientific notation.

6. Find the GCF of $15x^3y^2$ and $25x^2y$.

7. Solve the inequality and graph the solution on a number line. $8s - 8 \le 24$

8. Factor. $x^2 - 6x - 7$

9. Use the quadratic formula to solve for d. $d^2 + 2d - 3 = 0$

10. The area of a rectangle is 72 cm^2. If the length is 9 cm, what is the width?

11. Juan caught a fish that was 31 inches less than twice the length of his dad's fish. If his dad's fish was half a yard in length, how long was Juan's fish?

12. Solve for c. $11c + 36 = 8c$

13. $6\frac{2}{3} + 5\frac{1}{9} = ?$

14. Multiply. $(3x + 4)(4x + 5)$

15. $-73 - (-39) = ?$

16. Solve this system of equations. $\begin{aligned} x - 3y &= -3 \\ x + 3y &= 9 \end{aligned}$

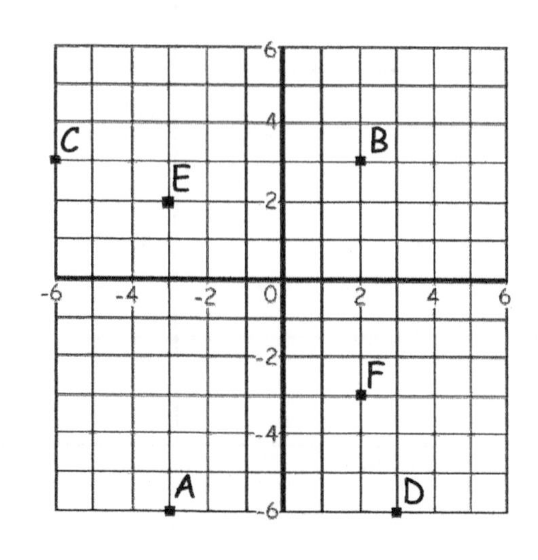

17. What number is 25% of 80?

18. Write $\frac{6}{25}$ as a decimal and as a percent.

19. Graph the linear equation. $y = \frac{3}{2}x + 1$

20. Find the coordinates of points D and E.

1.	2.	3.	4.
5.	6.	7.	8.
9.	10.	11.	12.
13.	14.	15.	16.
17.	18.	19.	20.

Lesson #89

1. Write the quadratic formula.

2. A line that includes points (2, 6) and (4, 9) has what slope?

3. Factor. $a^2 - 16$

4. Simplify. $7a^0b^{-3}$

5. Graph the solution on a number line. $y < 3$ or $y \geq -1$

6. How do you calculate the area of a triangle?

7. Solve for x. $\frac{2}{3}x = -12$

8. 80% of what number is 32?

9. $(-16)+(-24)+(18)=?$

10. Find the value of y. $18+3y=5y-4$

11. A seven sided polygon is a(n) _______________.

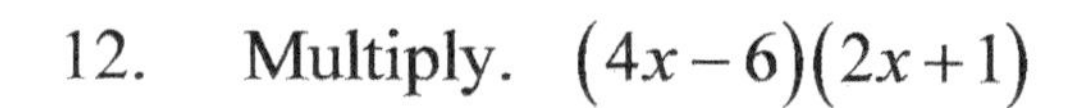

12. Multiply. $(4x-6)(2x+1)$

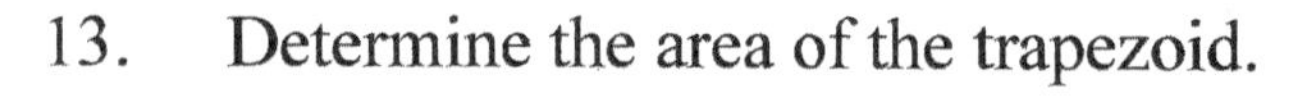

13. Determine the area of the trapezoid.

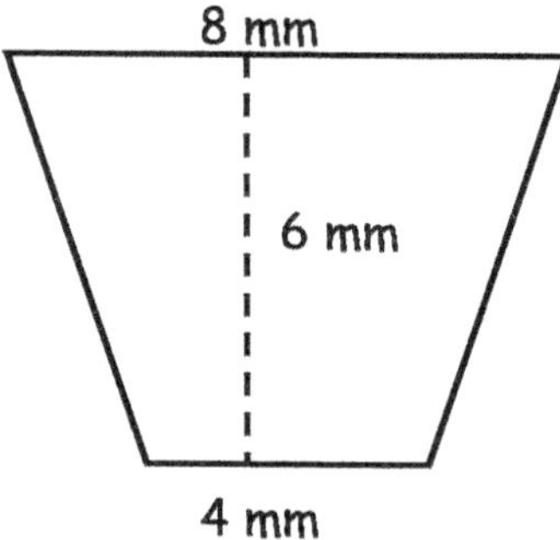

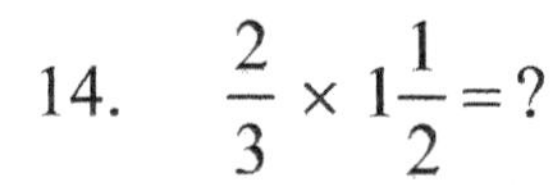

14. $\frac{2}{3} \times 1\frac{1}{2} = ?$

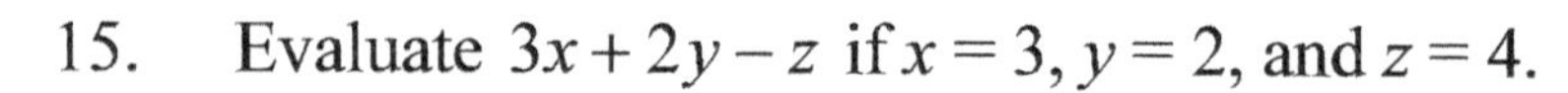

15. Evaluate $3x+2y-z$ if $x=3$, $y=2$, and $z=4$.

16. Translate into an algebraic phrase. *The quotient of a number and three.*

17. Write the formula for finding the volume of a cylinder.

18. $\frac{-56}{8} = ?$

19. When $x = \{0, -3, 2\}$, find the corresponding values for y in the equation $y = -4x$.

20. Find the difference.

$$\begin{array}{r} 10a^3 - 6a^2 + 4 \\ -\ \ 8a^3 + 2a^2 - 3 \\ \hline \end{array}$$

1.	2.	3.	4.
5.	6.	7.	8.
9.	10.	11.	12.
13.	14.	15.	16.
17.	18.	19.	20.

Lesson #90

1. $\frac{8}{9} \times \frac{12}{16} = ?$

2. Write the slope-intercept form of a linear equation.

3. $1.8 - 0.645 = ?$

4. Multiply. $(y-4)(y-10)$

5. Determine the value of $x^2 - 5$ when $x = 5$.

6. Simplify. $(c^5)^2$

7. A triangle with two sides congruent is a(n) __________ triangle.

8. Write 4.602×10^5 in standard notation.

9. Simplify. $9x^{-3}y^2z^{-2}$

10. Find the missing measurement, x.

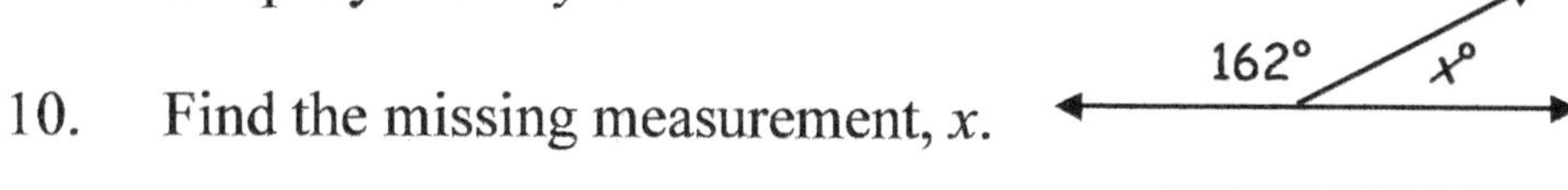

11. $-99 + (-76) = ?$

12. Find the area of the parallelogram.

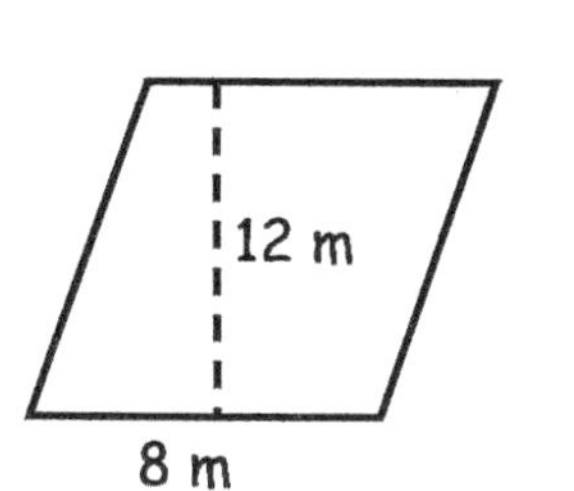

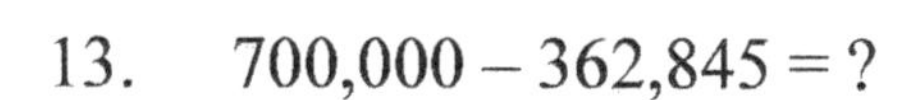

13. $700{,}000 - 362{,}845 = ?$

14. A picture was enlarged so that its width went from 16 inches to 24 inches. What was the percent increase of the picture's width?

15. Solve for y. $2(y+7) = 16$

16. Find the average of 65, 75, 85 and 95.

17. Solve the system. $\begin{aligned} 3x - 2y &= -12 \\ 5x + 4y &= 2 \end{aligned}$

18. What value of x makes the two sides equal?
$9x - 5 = 6x + 13$

19. James had 15 CD's. He gave $\frac{2}{3}$ of them to Kendra. How many CD's did he give to Kendra?

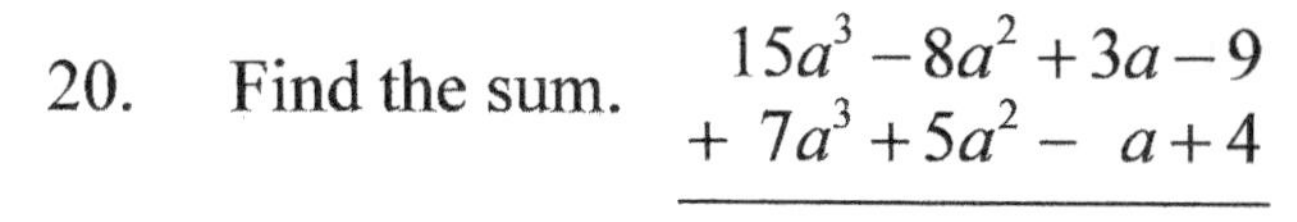

20. Find the sum.
$$\begin{array}{r} 15a^3 - 8a^2 + 3a - 9 \\ +\ 7a^3 + 5a^2 - \ a + 4 \\ \hline \end{array}$$

1.	2.	3.	4.
5.	6.	7.	8.
9.	10.	11.	12.
13.	14.	15.	16.
17.	18.	19.	20.

Lesson #91

1. Solve the inequality and graph it on a number line. $-3 < 2x + 5 \le 9$

2. $(-4)(-6)(-3) = ?$

3. Write 76% as a decimal and as a reduced fraction.

4. Write 0.00053 in scientific notation.

5. Multiply. $a^2\left(3a^2 - a + 2\right)$

6. $-|-17| = ?$

7. Solve using the quadratic formula. $x^2 - 4x + 3 = 0$

8. Factor. $x^2 - 10x + 9$

9. A line that includes points (5, –3) and (8, 3) has what slope?

10. Multiply. $\left(2c^3 - 3\right)^2$

11. If you flipped a coin twice and rolled a die once, what is the P(H, T, 6)?

12. $18 - 12\frac{5}{9} = ?$

13. $20 + 3\left[2(5-2) + 3\right] = ?$

14. What is the value of x? $\frac{4}{5}x - \frac{3}{5}x - 9 = 5$

15. $\begin{pmatrix} 3 & -4 & 0 \\ -6 & 5 & 2 \end{pmatrix} - \begin{pmatrix} -3 & 5 & -2 \\ 5 & -4 & 1 \end{pmatrix} = ?$

16. Solve for x. $\frac{7}{8} = \frac{x}{120}$

17. The Metroparks stocked Hollis Lake with 455 fish. If the park allows 27 fish to be caught each day, in how many days will the lake need to be restocked?

18. Solve the system. $\begin{aligned} x - 3y &= 9 \\ y &= 2 \end{aligned}$

19. Simplify. $\dfrac{3x^{-5}}{7y^{-2}}$

20. Simplify. $\dfrac{\left(2x^3\right)^2}{8x^5}$

1.	2.	3.	4.
5.	6.	7.	8.
9.	10.	11.	12.
13.	14.	15.	16.
17.	18.	19.	20.

Lesson #92

1. Solve for x. $3x+7=16+6x$

2. $56-(-12)=?$

3. Factor. $4x^2-12x-36$

4. $\frac{2}{9}$ of the 72 children in third grade are blonde. How many are not blonde?

5. Express the area of the rectangle as a polynomial in simplest form.

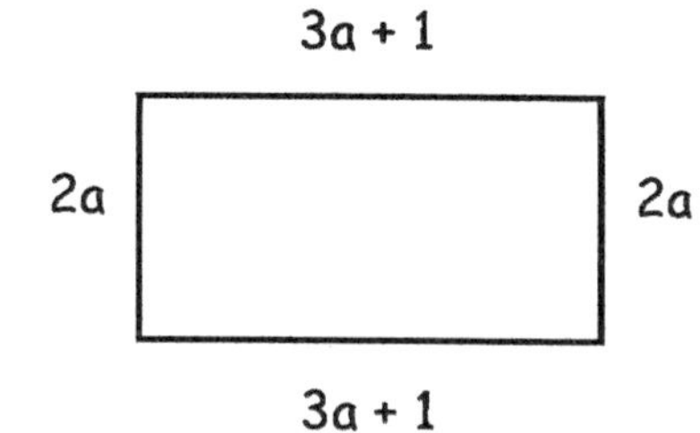

6. Multiply. $(3a+2b)(a-b)$

7. Solve for a. $\frac{4}{5}a=-60$

8. Write the quadratic formula.

9. A triangle with no congruent sides is a(n) ____________ triangle.

10. Write 6,200,000,000 in scientific notation.

11. What is the value of $\left(\frac{xyz}{5}+2x\right)$ when $x=3$, $y=5$, and $z=2$?

12. What is the formula for finding the area of a circle?

13. Write an algebraic expression for *six times a number decreased by eleven.*

14. Solve for x. $x-22=-58$

15. Give the slope and the y-intercept for the line with equation $2y=8$.

16. What are the corresponding values of y in the equation $y=-4x+2$ when $x=\{0,-3,2\}$?

17. Simplify. $\frac{6}{t^{-4}}$

18. What are the values of x and y? $3x+y=3$, $-3x+2y=-30$

19. Simplify. $\frac{(2a^2b)^3}{6ab^4}$

20. Find the sum. $(12x^3-6x^2-5x+2)+(8x^3+3x-8)$

1.	2.	3.	4.
5.	6.	7.	8.
9.	10.	11.	12.
13.	14.	15.	16.
17.	18.	19.	20.

Lesson #93

1. Simplify. $4\sqrt{25a^8b^{18}}$

2. $18 - 12\frac{3}{5} = ?$

3. Simplify. $\left(8a^2b^3\right)^2$

4. Factor. $25b^2 - 4$

5. What is the slope of a line passing through points (9, 4) and (5, 5)?

6. Solve for a. $-15 \le -3a + 12$

7. $16.25 + 8.476 + 3.005 = ?$

8. Find the circumference of a circle with a diameter of 9 mm.

9. Write 0.0301 in scientific notation.

10. Find the value of a in the equation. $11a + 8 = -2 + 9a$

11. Solve the system of equations. $\begin{aligned} 3x + y &= 7 \\ x - y &= 5 \end{aligned}$

12. Multiply. $4c\left(2c^2 - c + 3\right)$

13. What value of x makes the equation true? $-7x = 105$

14. A change from 17 pounds to 25 pounds represents what percent of increase?

15. $-66 - (29) = ?$

16. Find the area of the square.

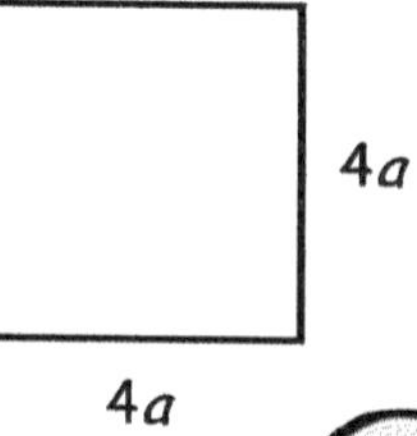

17. Solve for x. $\frac{x}{7} + 4 = 11$

18. Simplify. $\frac{6x^2y^{-3}}{z^{-6}}$

19. What percent of 80 is 56?

20. Nicole made 10 dozen cupcakes. She made 20% with chocolate icing, 30% with vanilla icing, and 50% with powdered sugar. How many of each kind of cupcake did she make?

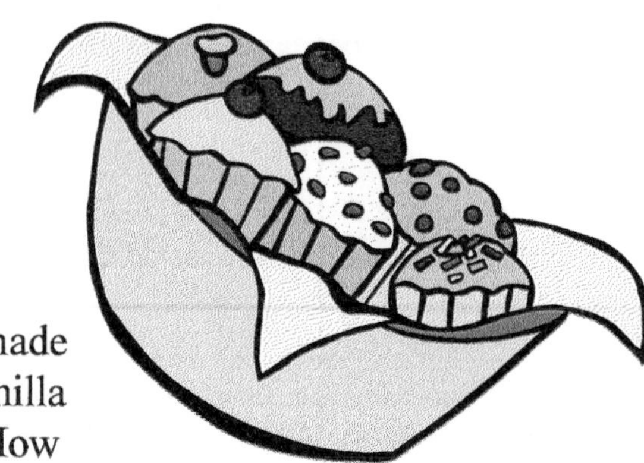

1.	2.	3.	4.
5.	6.	7.	8.
9.	10.	11.	12.
13.	14.	15.	16.
17.	18.	19.	20.

Lesson #94

1. Multiply. $(3c+4)(c+5)$

2. Raphael measured the diameter of his bicycle tire and found that it was 22 inches. What is the distance, to the nearest inch, around the tire?

3. Simplify. $-\sqrt{16y^{16}}$

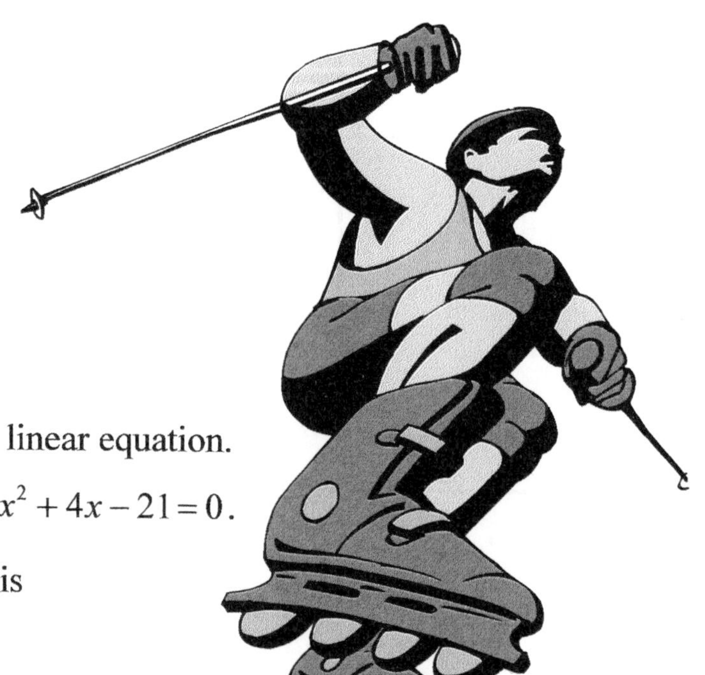

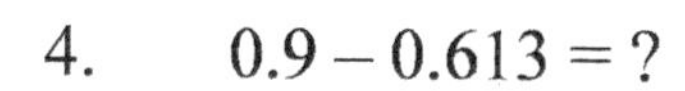

4. $0.9-0.613=?$

5. $1\frac{1}{2}\times\frac{2}{3}=?$

6. $\begin{pmatrix} 8 & 0 & -4 \\ 6 & -3 & 5 \end{pmatrix}+\begin{pmatrix} -6 & -5 & 7 \\ 2 & -6 & -1 \end{pmatrix}=?$

7. Write the slope-intercept form of a linear equation.

8. Use the quadratic formula to solve $x^2+4x-21=0$.

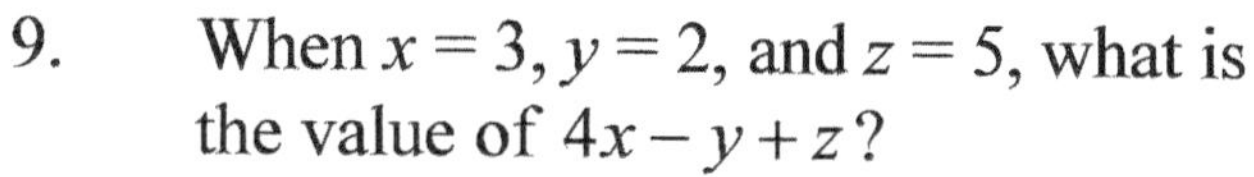

9. When $x=3$, $y=2$, and $z=5$, what is the value of $4x-y+z$?

10. $35\div5+2\cdot4+3-1=?$

11. Solve for y. $y-319=255$

12. Write an expression for *the product of a number and 7 divided by 3.*

13. Find the area of the square shown to the right.

x - 8

x - 8

14. $-76+(55)=?$

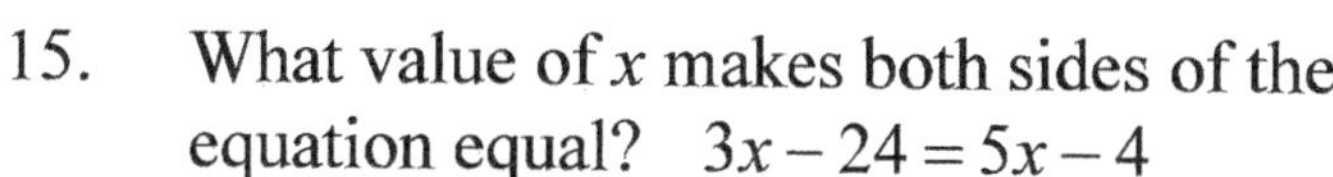

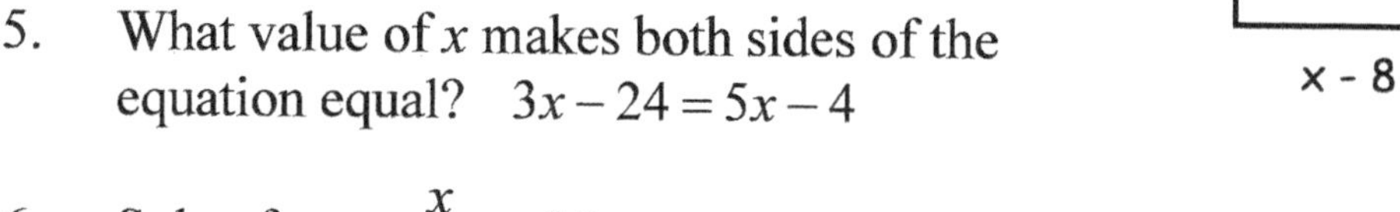

15. What value of x makes both sides of the equation equal? $3x-24=5x-4$

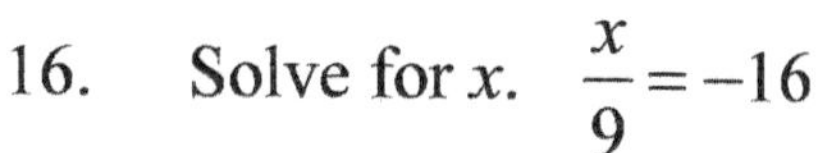

16. Solve for x. $\frac{x}{9}=-16$

17. Graph the solution to $11-3x\leq26$ on a number line.

18. Simplify. $\frac{12a^3 3b^2 c^5}{14abc^3}$

19. Simplify. 6^{-2}

20. Find the difference. $(9y^3+6y^2-4y+2)-(6y^3+2y-7)$

1.	2.	3.	4.
5.	6.	7.	8.
9.	10.	11.	12.
13.	14.	15.	16.
17.	18.	19.	20.

Lesson #95

1. Identify the slope and the y-intercept in the equation, $y = \frac{2}{5}x + 2$.

2. A ten-sided polygon is a(n) __________.

3. $60 - 8 \cdot 5 + 12 \div 2 = ?$

4. Simplify. $-\sqrt{16a^{20}}$

5. Which is greater, 80% or $\frac{17}{20}$?

6. Rachel's watering can is in the shape of a cylinder. If the radius is 3 inches and the height is 10 inches, what is the volume of the watering can?

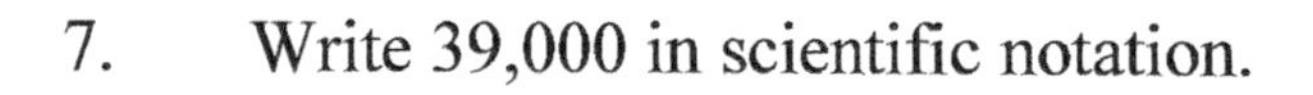

7. Write 39,000 in scientific notation.

8. Find the missing measurement, x.

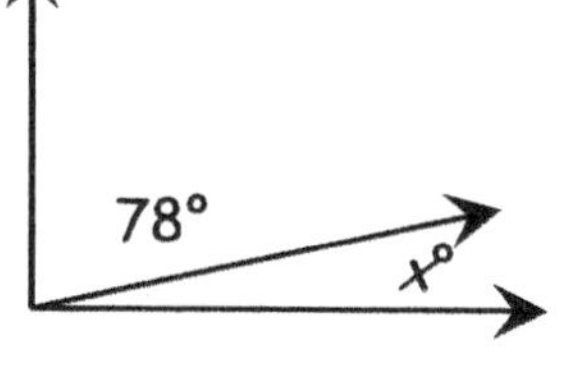

9. Solve for x. $3x - 8 = -17$

10. Simplify. $\left(7x^3y^2\right)^3$

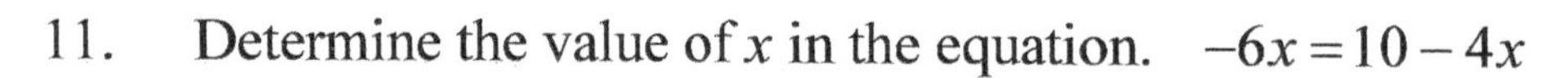

11. Determine the value of x in the equation. $-6x = 10 - 4x$

12. Simplify. $\frac{12x^{-2}y^3z^{-4}}{4w^{-4}}$

13. Factor out the GCF. $6x^2 - 21x - 12$

14. How many feet are in 12 yards?

15. Find the area of the triangle.

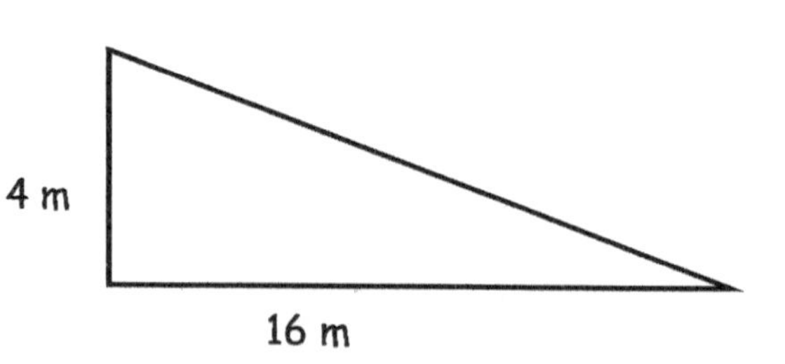

16. Write the quadratic formula.

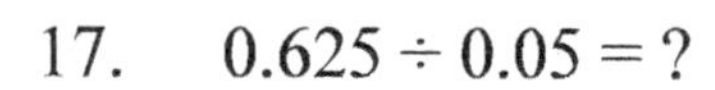

17. $0.625 \div 0.05 = ?$

18. $6\frac{1}{9} + 7\frac{2}{3} = ?$

19. Solve for x. $\frac{3}{8} = \frac{x}{104}$

20. Simplify. $\frac{\left(4a^5\right)\left(2a^2\right)}{2a^3}$

1.	2.	3.	4.
5.	6.	7.	8.
9.	10.	11.	12.
13.	14.	15.	16.
17.	18.	19.	20.

Lesson #96

1. Multiply. $2y^2(4y-5)$

2. Simplify. $\sqrt{3}\cdot\sqrt{51}$

3. $-7(-9)=?$

4. Solve for x. $\frac{3}{5}x=30$

5. Solve for d using the quadratic formula. $d^2-d-6=0$

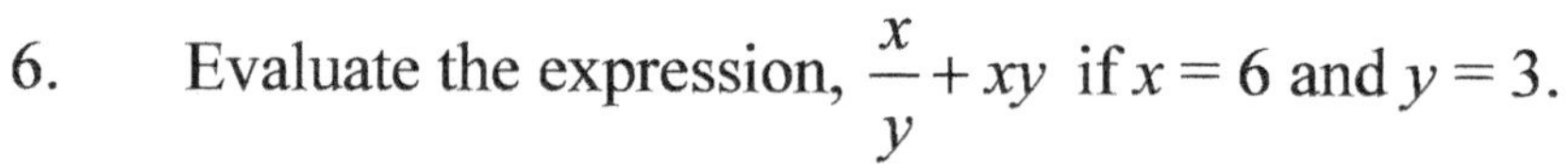

6. Evaluate the expression, $\frac{x}{y}+xy$ if $x=6$ and $y=3$.

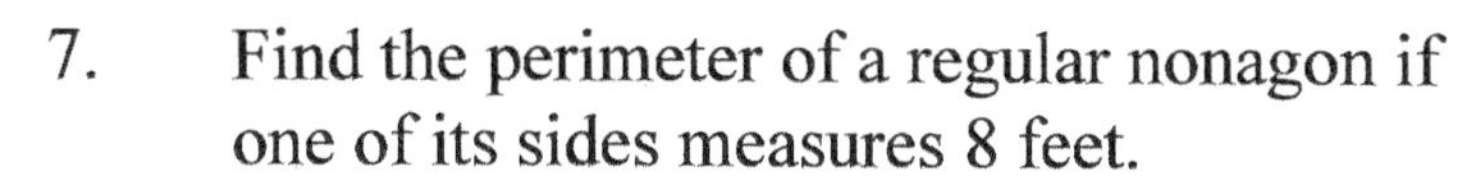

7. Find the perimeter of a regular nonagon if one of its sides measures 8 feet.

8. $2.35\times 0.4=?$

9. Find the LCM of $9a^3b^4c^2$ and $14a^2b^2c^2$.

10. $24\frac{1}{7}-18\frac{5}{7}=?$

11. Factor the trinomial. r^2+r-6

12. Solve this quadratic equation. $x^2-64=0$

13. $\begin{pmatrix}-2 & -7\\ 0 & 1\end{pmatrix}-\begin{pmatrix}-3 & 5\\ -2 & 6\end{pmatrix}=?$

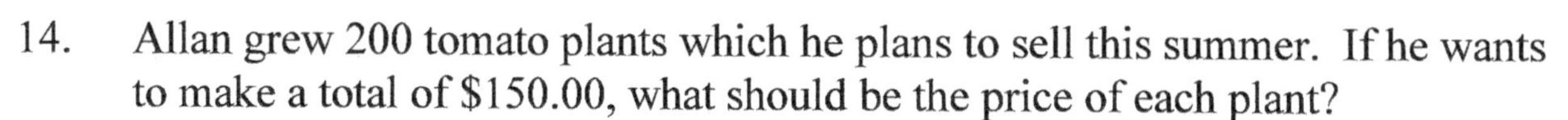

14. Allan grew 200 tomato plants which he plans to sell this summer. If he wants to make a total of $150.00, what should be the price of each plant?

15. Simplify. $\frac{6x^{-2}y^2}{4z^{-3}}$

16. Write $\frac{6}{25}$ as a decimal and as a percent.

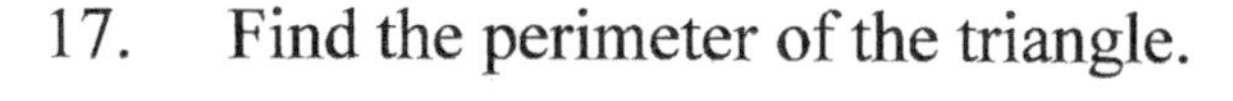

17. Find the perimeter of the triangle.

9a + 6

8a - 9

6a + 8

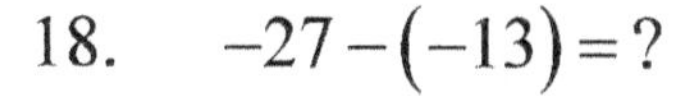

18. $-27-(-13)=?$

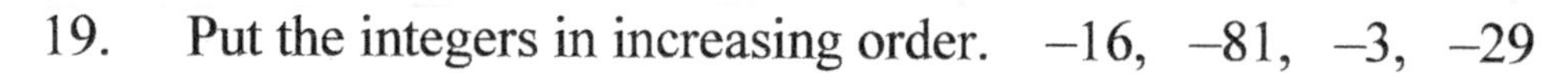

19. Put the integers in increasing order. $-16,\ -81,\ -3,\ -29$

20. Find the sum. $(12a^4-8a^3+6a^2+5)+(6a^4-3a^2+5a-2)$

1.	2.	3.	4.
5.	6.	7.	8.
9.	10.	11.	12.
13.	14.	15.	16.
17.	18.	19.	20.

Lesson #97

1. Simplify. $\sqrt{\frac{4}{9}}$

2. Write 0.00000725 in scientific notation.

3. Solve for y. $2y-6=12$

4. What number is 70% of 30?

5. Simplify. $(5x^2y)(-2xy^2)$

6. In his collection, Mario has a baseball card that his grandfather purchased for 25¢. The value of the card is now 5000% of the original value. What is the current value of the baseball card?

7. Solve using the quadratic formula. $3y^2-5y-4=0$
   Put irrational solutions in simplest radical form.

8. Factor. a^2-25

9. Solve for n. $8n-12=5n$

10. Simplify. $6x^{-4}y^{-3}z^2$

11. Identify the slope and the y-intercept for the linear equation, $y=\frac{4}{5}x+3$.

12. Solve for y. $y-155=214$

13. Solve the inequality. $x-1\geq-3$ or $4<2x-2$

14. Simplify. $\sqrt{9a^{14}}$

15. Factor. $x^2-2x-24$

16. A parallelogram with a base of 34 meters and a height of 5 meters has what area?

17. Evaluate $\frac{4x}{y}+xy$ when $x=4$ and $y=2$.

18. $14\frac{2}{3}+8\frac{1}{9}=?$

19. What are the values for y in $y=2x+1$ when $x=\{0,-4,3\}$?

20. $0.25\times0.5=?$

1.	2.	3.	4.
5.	6.	7.	8.
9.	10.	11.	12.
13.	14.	15.	16.
17.	18.	19.	20.

Lesson #98

1. Multiply. $5a\left(4a^2 - 3a + 7\right)$

2. $-32 - (+16) = ?$

3. Simplify. $14\sqrt{25y^{28}}$

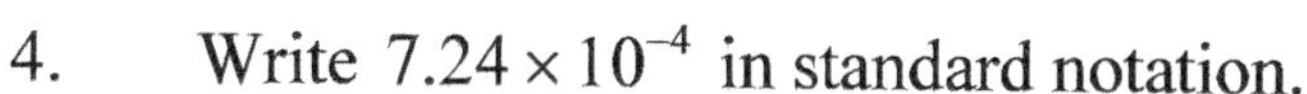

4. Write 7.24×10^{-4} in standard notation.

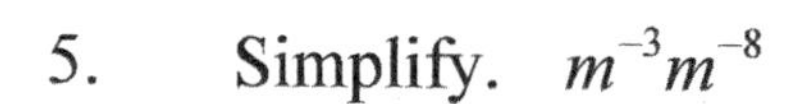

5. Simplify. $m^{-3}m^{-8}$

6. Multiply. $\left(5a^2 + 1\right)^2$

7. Factor the trinomial. $y^2 + 2y - 15$

8. $4.3 + 76.257 = ?$

9. What is the perimeter of a heptagon if the sides measure 5 inches each?

10. Find the area of the triangle.

4y

5y + 3

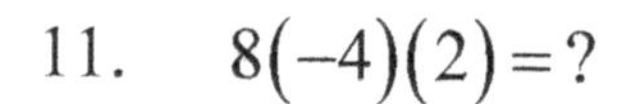

11. $8(-4)(2) = ?$

12. $-|-24| = ?$

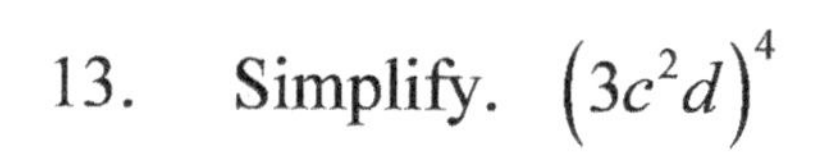

13. Simplify. $\left(3c^2d\right)^4$

14. $23\frac{1}{9} - 18\frac{5}{9} = ?$

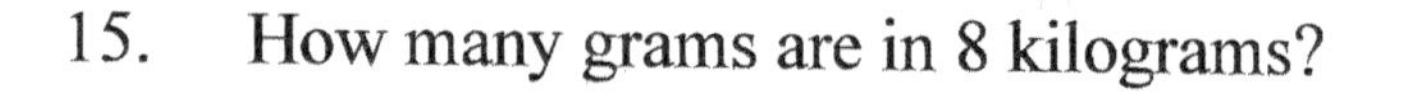

15. How many grams are in 8 kilograms?

16. Use the quadratic formula to solve. $c^2 + 6c + 8 = 0$

17. In the mall parking lot, the ratio of trucks to cars is 7 to 9. If there are 98 trucks, how many cars are in the parking lot?

18. $x + 6 < 2$ Solve the inequality and graph its solution on a number line.

19. $-62 \bigcirc -41$

20. Find the difference.

$$\begin{array}{r} 8a^4 - 5a^3 + 2a^2 + 9a - 7 \\ -\ \ 5a^4 \qquad\quad - a^2 + 5a + 3 \\ \hline \end{array}$$

1.	2.	3.	4.
5.	6.	7.	8.
9.	10.	11.	12.
13.	14.	15.	16.
17.	18.	19.	20.

Lesson #99

1. Multiply. $(2y+5)(3y+1)$

2. Simplify. $\sqrt{16a^{20}}$

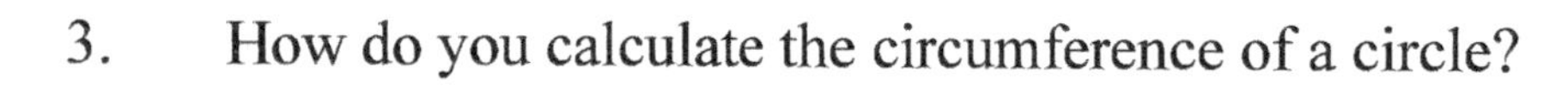

3. How do you calculate the circumference of a circle?

4. Write 2.8×10^5 in standard notation.

5. $0.008 \times 0.007 = ?$

6. Solve for x. $\frac{4}{5} = \frac{x}{75}$

7. $28 + (-13) + (-12) = ?$

8. How many cups are in 7 pints?

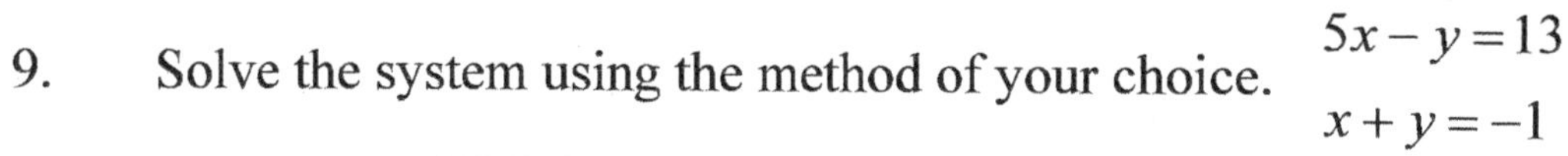

9. Solve the system using the method of your choice. $\begin{aligned} 5x - y &= 13 \\ x + y &= -1 \end{aligned}$

10. Simplify. $\sqrt{81m^{16}n^{10}}$

11. Calculate the volume of the cube. (6 in.)

12. Solve for x using the quadratic formula. $x^2 - x - 6 = 0$

13. Factor. $25x^3 - 15x^2 + 5x$

14. Determine the equation of the line through points (2, 5) and (–3, – 4).

15. Fran is baking cookies. The recipe she is making calls for $1\frac{1}{2}$ cups of butter. There are 4 sticks in one pound of butter, and each stick is equal to 8 tablespoons or $\frac{1}{2}$ cup. How many sticks of butter does Fran need? What is the weight, in pounds, of the butter she needs for her recipe?

16. Multiply. $2y(3y^3 - 6y + 1)$

17. 70% of 90 is what number?

18. Simplify. $-\sqrt{49a^{12}b^{10}}$

19. When $a = 6$ and $b = 2$, what is the value of $ab + \frac{a}{b}$?

20. Simplify. $\frac{3a^0b^{-1}c}{15bc^{-3}}$

1.	2.	3.	4.
5.	6.	7.	8.
9.	10.	11.	12.
13.	14.	15.	16.
17.	18.	19.	20.

Lesson #100

1. Find $\frac{4}{7}$ of 28.

2. Simplify. -2^{-4}

3. Write the slope-intercept form of a linear equation.

4. Factor. $6t^2 - t - 15$

5. Write 5,320,000,000 in scientific notation.

6. Multiply. $(7a+b)(3a-b)$

7. What is the P(3, H, 1, T) on 2 flips of a coin and 2 rolls of a die?

8. $80 - 24 \div 6 + 3 \cdot 3 - 2 = ?$

9. Simplify. $\left(3x^3y^2z^2\right)^3$

10. Round 87,346,297 to the nearest ten thousand.

11. What is the value of x in $5x - 3 = 10x + 7$?

12. Write an algebraic phrase for *a number decreased by 5 divided by 2.*

13. Find the difference. $\left(16a^3 - 8a^2 + 5a - 9\right) - \left(7a^3 + 2a^2 - 3a + 4\right)$

14. Solve for x. $2x - 8 = 16$

15. The farmer's truck weighs $2\frac{1}{2}$ tons when it is empty. The farmer loads the truck with $1\frac{1}{2}$ tons of potatoes and $\frac{1}{2}$ ton of carrots. How many tons does the loaded truck weigh?

16. Simplify. $\dfrac{p^{-4}y^5z}{z^{-5}y^3p^2}$

17. $\begin{pmatrix} 7 & -4 \\ -5 & 1 \end{pmatrix} + \begin{pmatrix} -5 & 2 \\ 3 & -4 \end{pmatrix} = ?$

18. Use substitution to solve the system. $\begin{array}{l} y = -3x + 2 \\ x = 3 \end{array}$

19. Simplify. $\sqrt{27}$

20. Find the area of the rectangle.

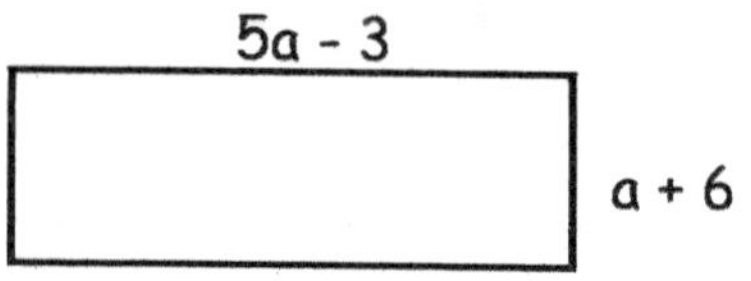

1.	2.	3.	4.
5.	6.	7.	8.
9.	10.	11.	12.
13.	14.	15.	16.
17.	18.	19.	20.

Lesson #101

1. Solve the inequality for a. $7a-9 \geq 35-4a$

2. Find the median, the mode, and the range of 73, 29, 15, 29 and 57.

3. $\dfrac{-264}{-4}=?$

4. What is the percent of increase from 18 to 32 yards? Round to the nearest whole number.

5. Simplify. $18\sqrt{2}-7\sqrt{2}+12\sqrt{2}$

6. Calculate the perimeter of the square.

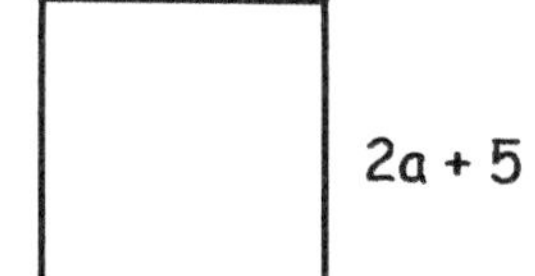

7. Write 5.3×10^6 in standard notation.

8. $85 \times 46 = ?$

9. What is the slope of the line that includes points (8, 2) and (6, 8)?

10. Multiply. $(2c-3)(3c-7)$

11. Find the LCM of $9x^3y^2$ and $14x^2y^2z^2$.

12. Simplify. $14\sqrt{7}+6\sqrt{7}$

13. A DVD player costs $249. The county sales tax is 8%. What will be the total cost of the DVD player?

14. Simplify. $\sqrt{50}$

15. Factor. $4x^2-12x-7$

16. Find the difference. $\begin{array}{r} 3y+8 \\ -\ 3y-9 \\ \hline \end{array}$

17. Simplify. $\dfrac{3ab(6a^2b^4)}{9a^3b^2}$

18. Simplify. $\dfrac{3x^{-5}}{7y^{-2}}$

19. $6\dfrac{2}{5}+9\dfrac{1}{4}=?$

20. Simplify. $5\sqrt{20}-6\sqrt{32}$

1.	2.	3.	4.
5.	6.	7.	8.
9.	10.	11.	12.
13.	14.	15.	16.
17.	18.	19.	20.

Lesson #102

1. Factor. $m^2 - 49$

2. Solve for a. $a + 57 = -124$

3. A circle has a diameter of 14 meters. Find the area of the circle.

4. Simplify. $7\sqrt{8} - \sqrt{18}$

5. When $x = \{-5, 0, 2\}$, what are the corresponding y-values in $y = -3x$?

6. Solve the system of equations. $\begin{array}{l} 2x + 9y = 4 \\ -5x - 9y = 17 \end{array}$

7. Identify the slope and the y-intercept. $y = 4x - 5$

8. Evaluate $\dfrac{abc}{4} + 2a$ when $a = 2$, $b = 5$, and $c = 4$.

9. Solve for x. $|5 - x| = 3$

10. Simplify. $\sqrt{4a^{10}b^{12}}$

11. Write 0.00000938 in scientific notation.

12. Multiply. $(x + 8)(x - 9)$

13. Write the quadratic formula.

14. Factor the trinomial. $a^2 - a - 42$

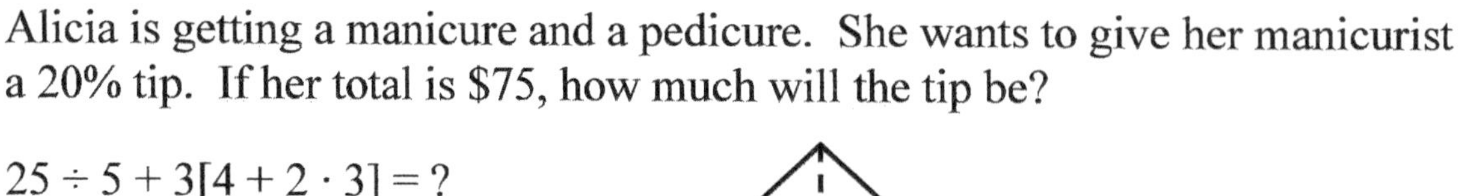

15. Alicia is getting a manicure and a pedicure. She wants to give her manicurist a 20% tip. If her total is \$75, how much will the tip be?

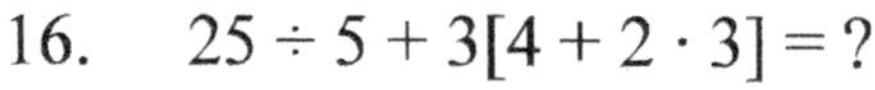

16. $25 \div 5 + 3[4 + 2 \cdot 3] = ?$

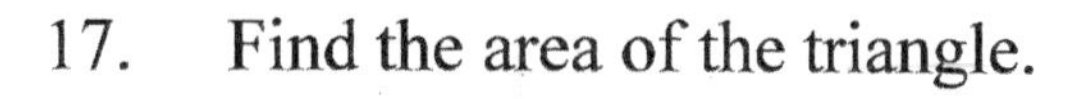

17. Find the area of the triangle.

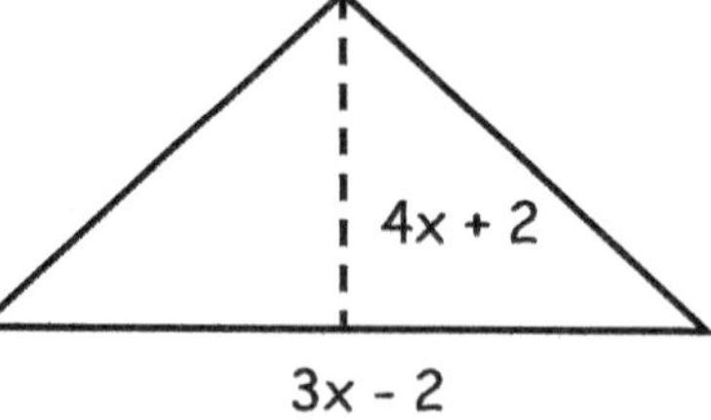

18. Simplify. $9x^3y^{-2}z$

19. Translate the phrase into an algebraic expression: *Twice a number divided by another number, plus four.*

20. What is the equation of the line passing through point (–1, 5) with a y-intercept of 3?

1.	2.	3.	4.
5.	6.	7.	8.
9.	10.	11.	12.
13.	14.	15.	16.
17.	18.	19.	20.

Lesson #103

1. Simplify. $\sqrt{10^4}$

2. $18 + 3[4 \cdot 2 + 1] = ?$

3. Solve for x. $|x-2|=3$

4. Simplify. $\sqrt{27}-\sqrt{18}$

5. What is the value of n in the equation, $14 + 3n = n - 14$?

6. Write 7.24×10^{-4} in standard notation.

7. Multiply. $-3b^4\left(2b^2-6b+2\right)$

8. Simplify. $\left(2\sqrt{18}\right)^2$

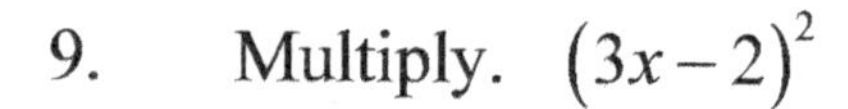

9. Multiply. $(3x-2)^2$

10. Solve for x. $\frac{x}{9}+14=21$

11. Find the area of the trapezoid.

14 m

9 m

6 m

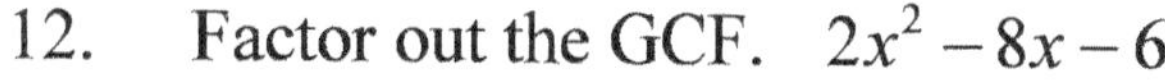

12. Factor out the GCF. $2x^2-8x-6$

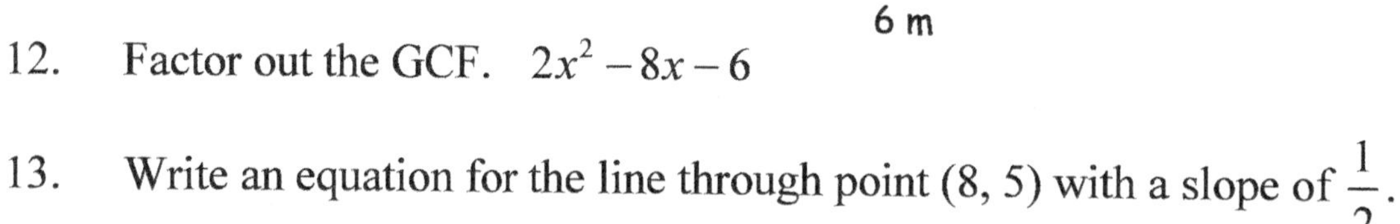

13. Write an equation for the line through point (8, 5) with a slope of $\frac{1}{2}$.

14. Write the formula for calculating the area of a circle.

15. A change from 1.4 m to 1.6 m represents an increase by what percent? Round to the nearest whole number.

16. What value of x makes the equation true? $x+146=-341$

17. How many feet are in 5 miles?

18. Write the quadratic formula.

19. Solve the system of equations. $\begin{matrix} 4x-9y=61 \\ 10x+3y=25 \end{matrix}$

20. Write $\frac{2}{50}$ as a reduced fraction and as a percent.

1.	2.	3.	4.
5.	6.	7.	8.
9.	10.	11.	12.
13.	14.	15.	16.
17.	18.	19.	20.

Lesson #104

1. Simplify. $-3x(-7x)(-2x^4)$

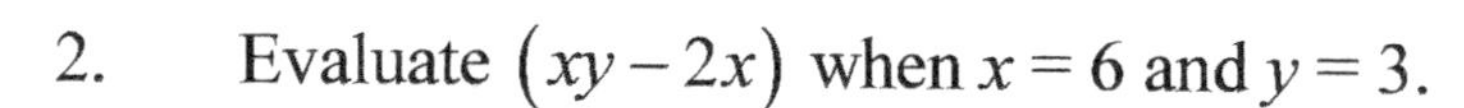

2. Evaluate $(xy-2x)$ when $x=6$ and $y=3$.

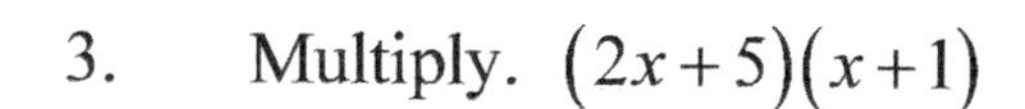

3. Multiply. $(2x+5)(x+1)$

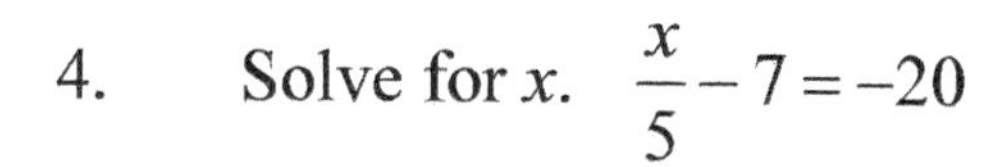

4. Solve for x. $\frac{x}{5}-7=-20$

5. $16-9\frac{4}{9}=?$

6. Solve for x. $-4x=35-9x$

7. Simplify. $-(2a^2bc^3)^2$

8. Factor. t^2-36

9. $-7(-9)(-2)=?$

10. Multiply. $(7x+y)(7x-y)$

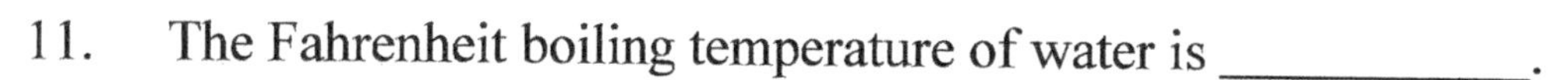

11. The Fahrenheit boiling temperature of water is __________.

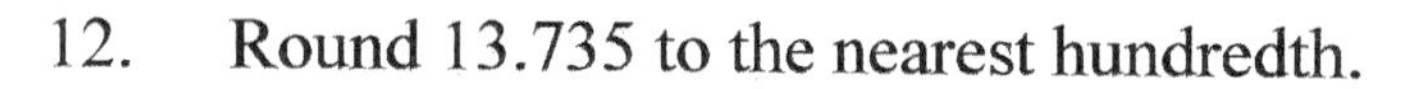

12. Round 13.735 to the nearest hundredth.

13. Simplify. $\left(\frac{3}{4}\right)^{-2}$

14. Write 86,000 in scientific notation.

15. The regular price for admission to the water park is $28 per person. There is a 25% discount off of the price of each ticket purchased today. How much will it cost for 4 people to go to the water park today?

16. Write the slope-intercept form of a linear equation.

17. Solve to find the value of x. $\frac{4}{5}x-\frac{3}{5}x-2=7$

18. The slope of a horizontal line is always ______________.

19. Simplify. $\frac{4xy^2(-3x^2y)}{10x^3y}$

20. Simplify. $\frac{6x^{-4}}{9y^{-3}}$

1.	2.	3.	4.
5.	6.	7.	8.
9.	10.	11.	12.
13.	14.	15.	16.
17.	18.	19.	20.

Lesson #105

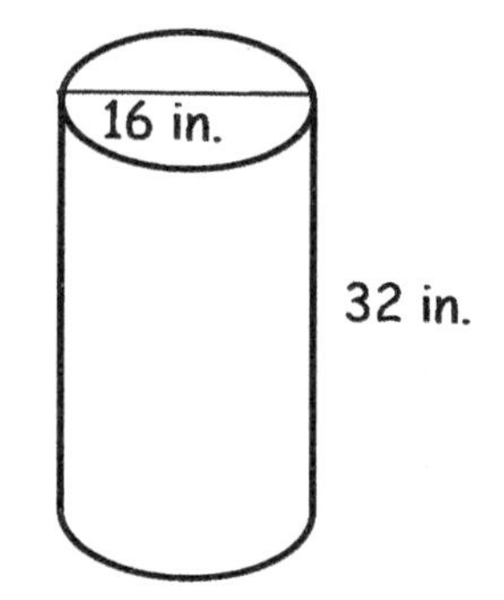

1. What is the volume of the cylinder?
2. Solve for x. $4 > -2 + x > -5$
3. $-83 + (-55) = ?$
4. Write 5.46×10^{-4} in standard notation.
5. Multiply. $4x(5x^2 + 3x - 2)$
6. Simplify. 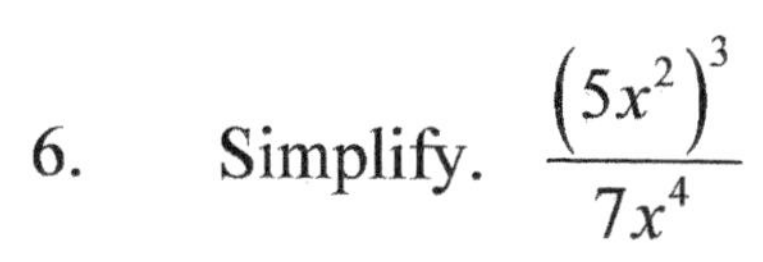 $\dfrac{(5x^2)^3}{7x^4}$
7. Determine the slope of a line passing through (2, 1) and (6, 8).
8. Multiply. $(x^{-8}y^{-5})(x^{10}y^{-2})$
9. Find $\dfrac{3}{5}$ of 45.
10. Find the perimeter of the rectangle.

3a + 1

2a

11. 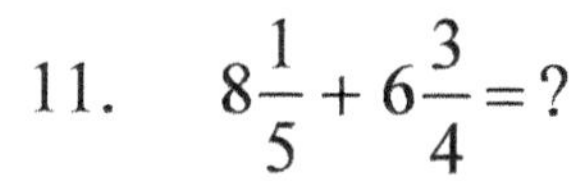 $8\frac{1}{5} + 6\frac{3}{4} = ?$
12. Multiply. 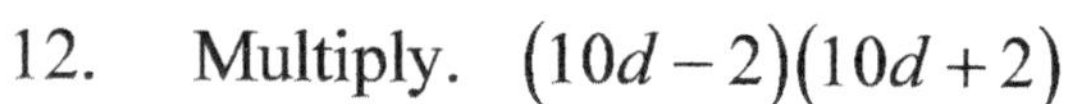 $(10d - 2)(10d + 2)$
13. 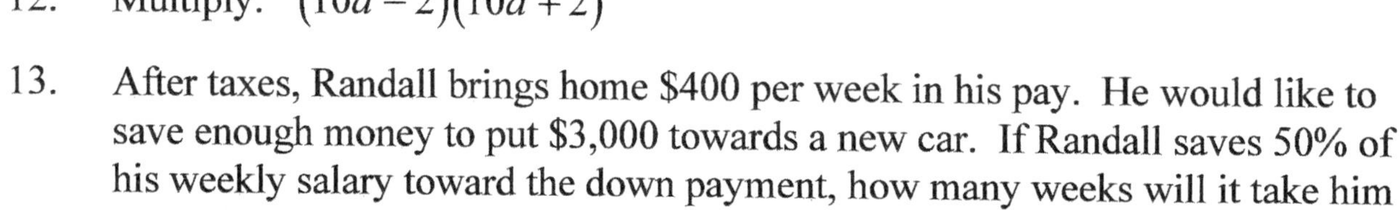 After taxes, Randall brings home $400 per week in his pay. He would like to save enough money to put $3,000 towards a new car. If Randall saves 50% of his weekly salary toward the down payment, how many weeks will it take him to reach his goal?
14. Simplify. $5\sqrt{81b^{30}}$
15. Simplify. $9\sqrt{2} - 3\sqrt{2}$
16. Solve the system of equations. $\begin{aligned} 2x + 7y &= 16 \\ -2x + 8y &= 14 \end{aligned}$

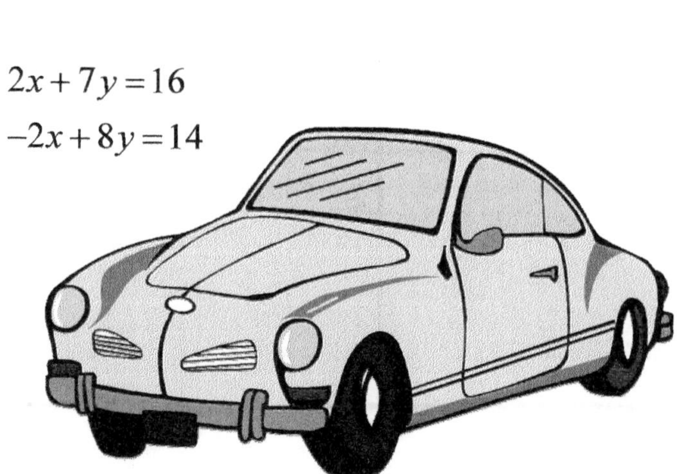

17. Simplify. $4\sqrt{75} + 6\sqrt{27}$
18. Simplify. $\dfrac{-4a^2b^3c}{-28abc}$
19. Simplify. $\dfrac{(2y)^0}{(3y)^0}$
20. Find the sum. 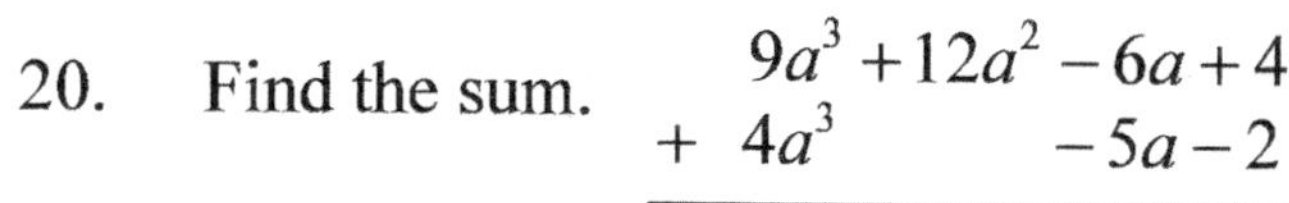

$$\begin{array}{r} 9a^3 + 12a^2 - 6a + 4 \\ +\ 4a^3 \qquad\quad - 5a - 2 \\ \hline \end{array}$$

1.	2.	3.	4.
5.	6.	7.	8.
9.	10.	11.	12.
13.	14.	15.	16.
17.	18.	19.	20.

Lesson #106

1. Simplify. $\left(4x^2y^3z^4\right)^2$

2. $30 + 3 \cdot 5 + 10 \div 2 - 1 = ?$

3. Simplify. $8a^{-3}b^2c^{-1}$

4. Solve for x. $-4x = 64$

5. Find the area of a triangle if the base is 54 cm and the height is 6 cm.

6. Water freezes at ________ °F.

7. Simplify. $3\sqrt{25xy} + 4\sqrt{36xy} - 2\sqrt{81xy}$

8. $18\frac{2}{7} - 8\frac{5}{7} = ?$

9. Evaluate $4a(a+b)$ when $a = 5$ and $b = 2$.

10. $\frac{-50}{-2} = ?$

11. Simplify. $-10\sqrt{49y^{44}}$

12. A triangle with two congruent sides is a(n) ______________ triangle.

13. $h + 5 \leq 16$ Graph the solution on a number line.

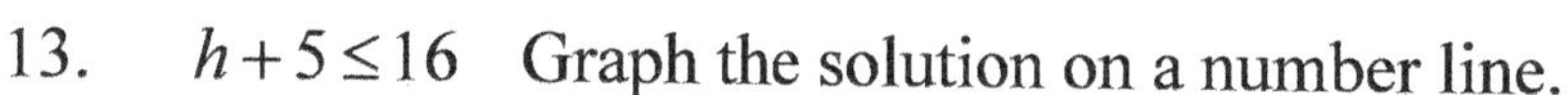

14. Write 0.0000413 in scientific notation.

15. Factor. $9k^2 - 64$

16. $0.56 \times 0.04 = ?$

17. Simplify. $-3\sqrt{28x^3y^7}$

18. What number is 60% of 40?

19. $\begin{pmatrix} 6 & -3 & 2 \\ 5 & 4 & -7 \end{pmatrix} - \begin{pmatrix} 5 & -2 & 0 \\ -3 & 2 & -5 \end{pmatrix} = ?$

20. Calculate the surface area of the rectangular prism.

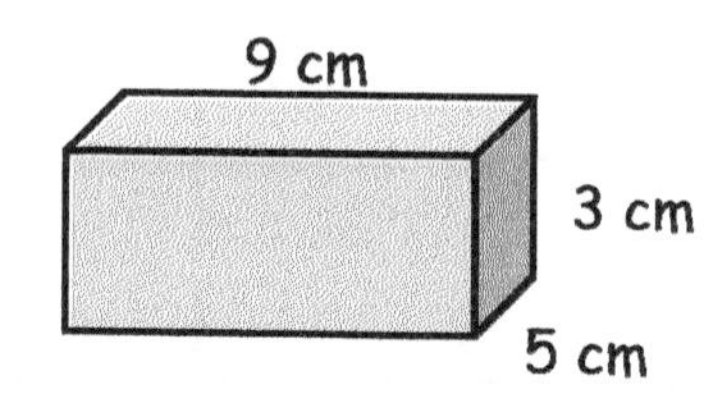

1.	2.	3.	4.
5.	6.	7.	8.
9.	10.	11.	12.
13.	14.	15.	16.
17.	18.	19.	20.

Lesson #107

1. Simplify. $\left(r^3s^2\right)^6$

2. A nine-sided polygon is called a(n) ____________.

3. Write 4,960,000,000 in scientific notation.

4. Multiply. $3a(2a+1)$

5. Simplify. $\left(2x^4z\right)\left(-3xy^3z\right)$

6. The slope of a vertical line is _________.

7. Find the area of a circle whose radius is 5 inches.

8. Solve for a. $a+46=92$

9. Simplify. $\sqrt{18}+\sqrt{3}$

10. Multiply. $\left(x^2-2\right)\left(x^2+5\right)$

11. Simplify. $\sqrt{25x^4y^6}$

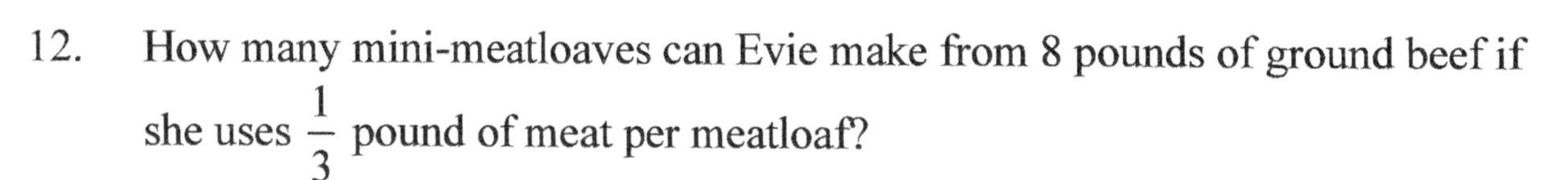

12. How many mini-meatloaves can Evie make from 8 pounds of ground beef if she uses $\frac{1}{3}$ pound of meat per meatloaf?

13. Write the quadratic formula.

14. Solve for x. $\frac{x}{13}=-7$

15. $\frac{7}{9}\times\frac{18}{28}=?$

16. Find the area of the square.

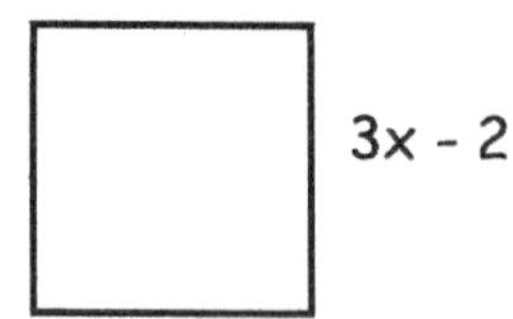

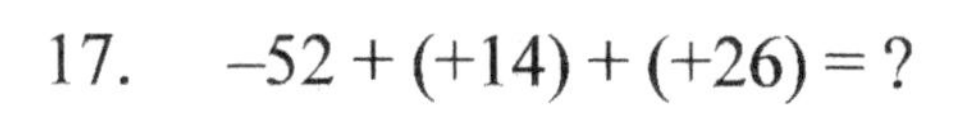

17. $-52+(+14)+(+26)=?$

18. Find the y-values in the equation, $y=-5x+1$ when $x=\{0, -3, 2)$.

19. Factor. $y^2+17y+72$

20. Find the difference.

$$\begin{array}{r} 15x^3-9x^2+4x-3 \\ -\quad 8x^3 \qquad\quad -2x+1 \\ \hline \end{array}$$

1.	2.	3.	4.
5.	6.	7.	8.
9.	10.	11.	12.
13.	14.	15.	16.
17.	18.	19.	20.

Lesson #108

1. Simplify. $\sqrt{2}\left(5-\sqrt{8}\right)$

2. Factor. $y^2+2y-15$

3. Multiply. $4x\left(3x^2-5x+2\right)$

4. Simplify. $6^{-2}a^2b^{-3}c^{-2}$

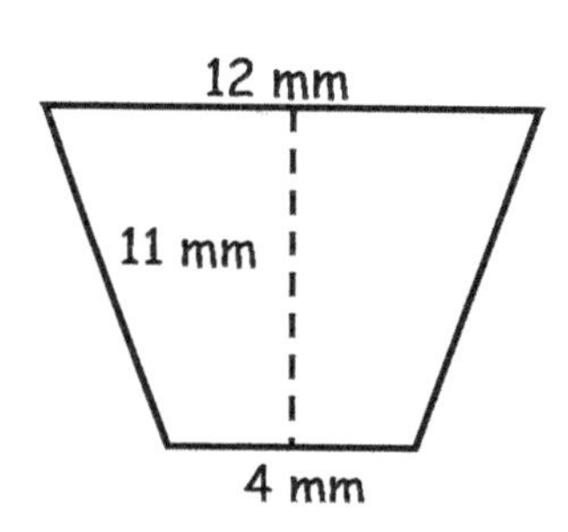

5. Solve for x. $7x-9=3x+19$

6. Find the area of the trapezoid.

7. Solve for a. $2a+12=24$

8. Simplify. $\left(4a^3b^2c\right)^2$

9. $86+(-25)=?$

10. Simplify. $\sqrt{3}\left(\sqrt{15}+\sqrt{4}\right)$

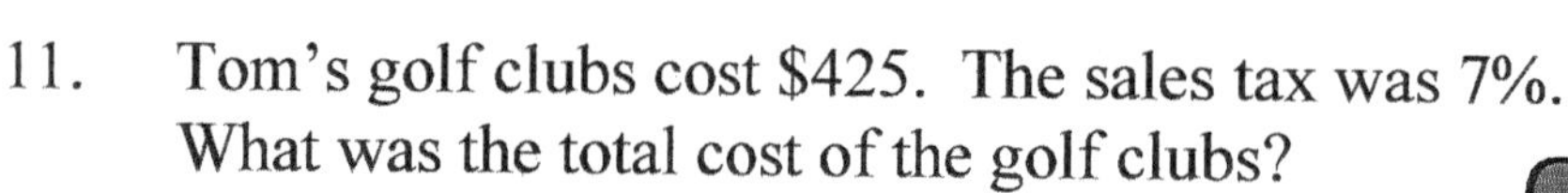

11. Tom's golf clubs cost \$425. The sales tax was 7%. What was the total cost of the golf clubs?

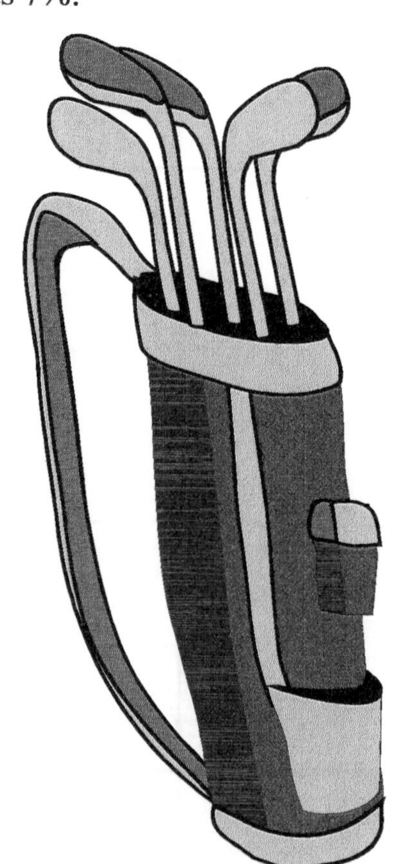

12. $800{,}000-396{,}278=?$

13. Simplify. $8\sqrt{9x^{14}y^{40}}$

14. Write 0.00084 in scientific notation.

15. Solve the system of equations. $\begin{aligned} x+4y&=14 \\ 6x-2y&=6 \end{aligned}$

16. Solve for x. $\frac{2}{-3}x=16$

17. What is the value of x in $\frac{7}{8}=\frac{x}{120}$?

18. Put these decimals in decreasing order.

    2.56 2.6 2.05 2.056

19. Solve the inequality for x. $3x-8\geq 7$

20. Find the sum. $\left(16x^3-8x^2-6x+5\right)+\left(4x^3+3x-2\right)$

1.	2.	3.	4.
5.	6.	7.	8.
9.	10.	11.	12.
13.	14.	15.	16.
17.	18.	19.	20.

Lesson #109

1. Factor. $m^2 - 10m + 25$
2. Write 0.00043 in scientific notation.
3. Find the slope of a line passing through points (3, 4) and (6, 2).
4. $-75 + (-43) = ?$
5. Find the perimeter of the square. 3x - 1
6. Multiply. $3b^4(2b^2 - 6b + 2)$
7. Solve for c. $c - 7 < -2$
8. Simplify. $\sqrt{81b^{17}}$
9. $1.8 + 5.64 = ?$
10. $x + 9 = -12$
11. A starting salary for a lab technician is $636 per week, while a senior lab technician earns $2,025 weekly. What is the percent of increase given for experience on the job?
12. Multiply. $(2y^2 - 3y + 1)(7y + 2)$
13. Simplify. $\frac{-2x^2y}{14xy^4}$
14. Simplify. $2a^{-3}b^2c^{-5}$
15. Solve for x. $\frac{3}{4}x = 24$
16. Simplify. $(2x^2)(5x^3)$
17. Solve using the quadratic formula. $a^2 - 7a + 10 = 0$
18. Solve for c. $3c + 7 = 6c + 16$
19. Multiply. $(x - 6)(x + 2)$
20. Find the median of these temperatures: 97°, 94°, 86°, 89°, 92° and 90°.

1.	2.	3.	4.
5.	6.	7.	8.
9.	10.	11.	12.
13.	14.	15.	16.
17.	18.	19.	20.

Lesson #110

1. Find the percent of change from 16 inches to 20 inches.

2. Write 65% as a decimal and as a reduced fraction.

3. $93-(-28)=?$

4. Evaluate $3x-y$ if $x=4$ and $y=2$.

5. Write an expression that represents *four more than a number.*

6. What is the value of x? $-4x=64$

7. Solve for x. $3x-3\leq 9$

8. Write the quadratic formula.

9. Simplify. $\left(5a^2b^3\right)^3$

10. $36\frac{3}{5}-12\frac{4}{5}=?$

11. Write 7.4×10^{-5} in standard notation.

12. Multiply. $\left(4x^2+5x-3\right)\left(x+2\right)$

13. 90% of what number is 36?

14. Simplify. $\frac{3x^{-5}}{7y^{-2}}$

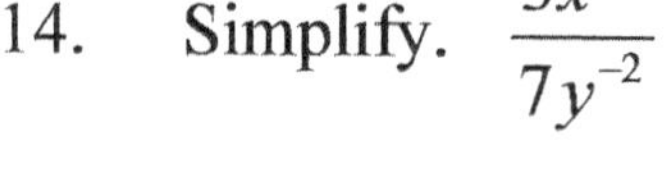

15. Find the values for y in the equation $y=2x-4$ when $x=\{-4, 2, 0\}$.

16. $\frac{5}{7}\cdot\frac{14}{25}=?$

17. $\begin{pmatrix}6 & -3\\0 & 2\end{pmatrix}+\begin{pmatrix}5 & -5\\-3 & -4\end{pmatrix}=?$

18. A triangle with no congruent sides is called a(n) ___________ triangle.

19. Solve the system of equations. $\begin{aligned}3x-2y&=6\\5x+7y&=41\end{aligned}$

20. Find the difference. $\begin{array}{r}9a^3+6a^2-4a+5\\-\;\;6a^3\qquad\;\;+2a-9\\\hline\end{array}$

1.	2.	3.	4.
5.	6.	7.	8.
9.	10.	11.	12.
13.	14.	15.	16.
17.	18.	19.	20.

Lesson #111

1. Factor. $5x^2 - 10x$
2. Write 0.0000062 in scientific notation.
3. Simplify. $\dfrac{h^{-8}}{h^{-3}}$
4. 25% of 60 is what number?
5. Find the product. $8c\left(c^2 - 5c + 3\right)$
6. Solve for t. $10t + 6 = 8t + 12$
7. Solve the proportion for x. $\dfrac{7}{9} = \dfrac{x}{135}$
8. Multiply. $(3y + 5)(3y - 5)$
9. $-62 + (-37) = ?$
10. Solve for a. $2a - 5 \geq a + 3$
11. Find the slope of the line through points (2, 5) and (6, 8).

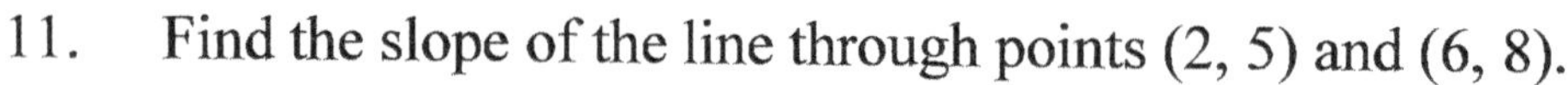

12. Simplify. $\sqrt{49a^{10}}$
13. Factor. $b^2 - 6b - 27$
14. How many quarts are in 12 gallons?
15. Find the percent of change from 8 feet to 14 feet.
16. A bag contains 8 red marbles and 10 white marbles. What is the probability of picking (red, red) if you replace the marble before making the second pick?
17. Find the area of a parallelogram whose base is 17 centimeters and whose height is 4 centimeters.
18. Factor. $a^2 - 25$
19. Simplify. $5x^2y\sqrt{18x^5y^7}$
20. Find the sum. $\begin{array}{r} 13x^2 - 8x + 9 \\ +\ 5x^2 \qquad - 3 \\ \hline \end{array}$

1.	2.	3.	4.
5.	6.	7.	8.
9.	10.	11.	12.
13.	14.	15.	16.
17.	18.	19.	20.

Lesson #112

1. Evaluate $\frac{ab}{3}+c$ if $a=6$, $b=9$, and $c=4$.

2. Solve. $42 \div 6+3 \cdot 3-2$

3. Multiply. $(x-7)(x+3)$

4. Simplify. $\sqrt{36a^{10}b^{14}}$

5. Write 3.4×10^{-3} in standard notation.

6. Solve for y. $9y-18=3y$

7. Find the GCF of $12x^3y^2z$ and $18x^2yz^3$.

8. Simplify. $c^4d^3c^{-2}$

9. Solve for x. $4x-5 \leq 19$

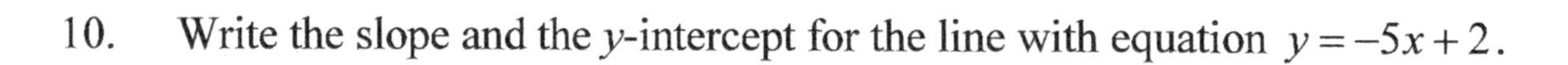

10. Write the slope and the y-intercept for the line with equation $y=-5x+2$.

11. What is the P(1, 3, 6) on three rolls of a die?

12. Factor. $a^2+15a+54$

13. Write the formula for finding the surface area of a rectangular prism.

14. Simplify. $2\sqrt{12a^9}$

15. $63-(-25)=?$

16. Simplify. $5\sqrt{3}+7\sqrt{3}$

17. Use the quadratic formula to solve. $a^2-9a+14=0$

18. Solve for c. $c+24=-88$

19. $-16+(-12)+14=?$

20. Simplify. $\frac{\sqrt{35}}{\sqrt{5}}$

1.	2.	3.	4.
5.	6.	7.	8.
9.	10.	11.	12.
13.	14.	15.	16.
17.	18.	19.	20.

Lesson #113

1. Solve for y. $4y = -20$

2. Write 850,000,000 in scientific notation.

3. Multiply. $(3d-7)(3d+7)$

4. What number is 80% of 30?

5. Solve for h. $h - 12 = -60$

6. Factor. $y^2 - 18y + 17$

7. $16\frac{2}{5} - 12\frac{4}{5} = ?$

8. Find the area of the square. 2x + 2

9. Find the values for y in the equation $y = -3x$ when $x = \{0, -2, 3\}$.

10. Simplify. $(8a^2b^3)^3$

11. $-33 + (-57) = ?$

12. Find the area of a circle if its radius is 5 meters long.

13. What is the quadratic formula?

14. Solve for a. $\frac{4}{5}a = 20$

15. Write $\frac{1}{4}$ as a decimal and as a percent.

16. Multiply. $(3y^2 + 5y - 4)(y + 3)$

17. Simplify. $\sqrt{3} \cdot 2\sqrt{7}$

18. Solve for x. $\frac{3}{7} = \frac{x}{91}$

19. Simplify. $\frac{3a^{-1}b^2c}{9a^2bc^{-2}}$

20. Solve this system of equations using any method. $x - y = 7$, $x + y = 39$

1.	2.	3.	4.
5.	6.	7.	8.
9.	10.	11.	12.
13.	14.	15.	16.
17.	18.	19.	20.

Lesson #114

1. Simplify. $3\sqrt{4}\cdot 5\sqrt{5}$

2. 25% of 60 is what number?

3. Factor. k^2-81

4. $(-6)(2)(-3)=?$

5. Solve using the quadratic formula. $a^2-9a+8=0$

6. Write the formula for finding the area of a triangle.

7. Multiply. $(3x+2)(2x+3)$

8. Solve for x. $\frac{2}{9}=\frac{x}{108}$

9. $\frac{4}{5}\cdot\frac{10}{12}=?$

10. Solve for a. $5(a+2)=30$

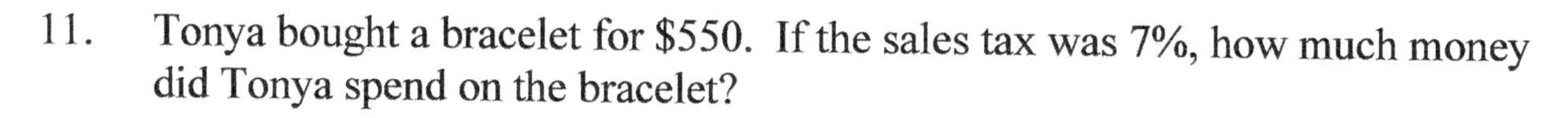

11. Tonya bought a bracelet for $550. If the sales tax was 7%, how much money did Tonya spend on the bracelet?

12. Write 3.23×10^4 in standard notation.

13. Simplify. $8x^{-3}y^2z^{-5}$

14. Which is greater, $\frac{6}{25}$ or 25%?

15. How many centimeters are in 17 meters?

16. Put these integers in decreasing order. −73, −16, 0, −3, −28

17. Write the slope of a line that passes through points (1, 3) and (6, 8).

18. Factor. $3x^3+9x^2-6x$

19. Simplify. $7\sqrt{64a^{20}}$

20. Find the difference.
$$\begin{array}{r} 9x^2-3 \\ -\ \ 2x^2+1 \\ \hline \end{array}$$

1.	2.	3.	4.
5.	6.	7.	8.
9.	10.	11.	12.
13.	14.	15.	16.
17.	18.	19.	20.

Lesson #115

1. Solve for y. $4y+2=5y+4$

2. Write 513,000 in scientific notation.

3. Solve for x. $\frac{x}{5}+7=12$

4. Factor out the GCF. $8a^3b+12a^4b^3-4ab^2$

5. Multiply. $\left(4x^2+2x+3\right)\left(2x+5\right)$

6. Write the slope-intercept form of a linear equation.

7. Simplify. $8n^4\cdot 2n^5$

8. Round 18.247 to the nearest tenth.

9. Evaluate. $30+6\left(3+2\cdot 4+1\right)$

10. Factor. $m^2-6m-40$

11. Write 0.44 as a percent and as a reduced fraction.

12. $19\frac{2}{7}-12\frac{6}{7}=?$

13. Evaluate $4xy+xy$ if $x=2$ and $y=3$.

14. $88+(-43)=?$

15. $\begin{pmatrix} 8 & 0 & -2 \\ 5 & 3 & 4 \end{pmatrix}+\begin{pmatrix} 2 & -5 & -4 \\ 8 & -1 & 2 \end{pmatrix}=?$

16. Graph the solution on a number line. $2h+6\le 18$

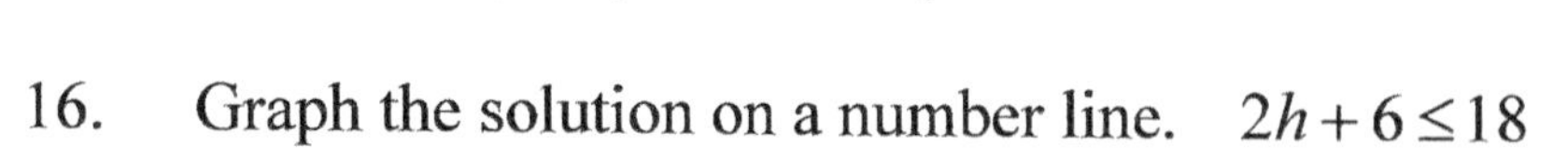

17. Simplify. $\frac{c^{-7}d^3}{c^3d}$

18. What is the percent of change from 19 yards to 25 yards? Round your answer to the nearest whole number.

19. Calculate the circumference of a circle whose radius is 4.5 feet.

20. Multiply. $6a\left(4a^2-3a+2\right)$

1.	2.	3.	4.
5.	6.	7.	8.
9.	10.	11.	12.
13.	14.	15.	16.
17.	18.	19.	20.

Lesson #116

1. Multiply. $(x+8)(x-3)$
2. What is 70% of 30?
3. $41+(-18)=?$
4. Solve for c. $c-26=-70$
5. Solve the system of equations using any method. $x-3y=-3$, $x+3y=9$
6. $3.62 \times 0.04 = ?$
7. Find the perimeter of the rectangle.

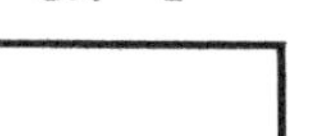

8. Solve for x. $-7x=91$
9. Write 8,000,000,000 in scientific notation.
10. Simplify. $3\sqrt{2}\cdot 7\sqrt{2}$
11. Solve for x. $\frac{5}{6}x-\frac{4}{6}x+3=12$
12. Simplify. $\sqrt{100a^{12}b^{16}}$
13. Use the quadratic formula to solve $a^2+2a-3=0$.
14. Simplify. $3\sqrt{2x^5}\cdot 4\sqrt{8x}$
15. The slope of a horizontal line is _____.

16. Solve for x. $\frac{5}{8}=\frac{x}{112}$
17. Solve for x. $3x-5=16$
18. $35-28\frac{5}{8}=?$
19. Solve for x. $\frac{-3}{5}x=15$
20. What are the coordinates for points A, B, and C?

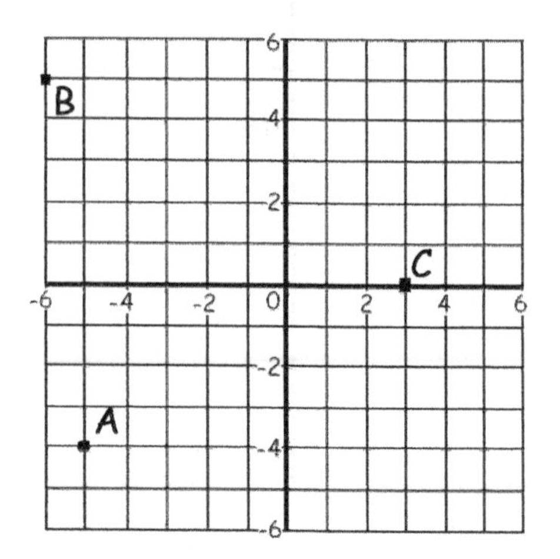

1.	2.	3.	4.
5.	6.	7.	8.
9.	10.	11.	12.
13.	14.	15.	16.
17.	18.	19.	20.

Lesson #117

1. $27+(-13)+42=?$

2. Write 0.000072 in scientific notation.

3. Simplify. $\left(8c^4d^5\right)^2$

4. Write the equation for the line through point (4, 6) having a slope of –5.

5. Factor. $2x^2+5x-3$

6. Find the area of the quadrilateral.

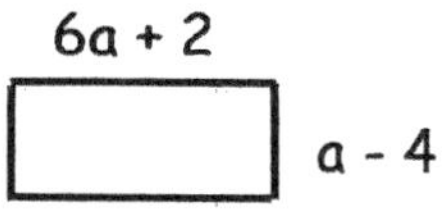

7. Simplify. $\dfrac{\sqrt{45}}{\sqrt{5}}$

8. $-97-(-38)=?$

9. How many feet are in 8 miles?

10. Simplify. $9\sqrt{5}-4\sqrt{5}$

11. What percent of 80 is 72?

12. Simplify. $\dfrac{(a+3)(a+4)}{(a-3)(a+4)}$

13. Find the missing measurement, x.

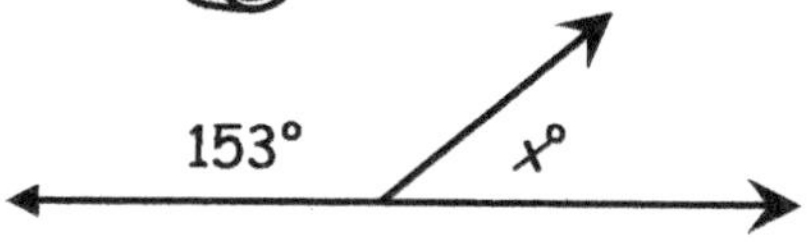

14. Factor out the GCF. $10a^3-5a^2+15a$

15. Solve the system using any method. $\begin{aligned}3x-y&=4\\x+5y&=-4\end{aligned}$

16. What is the P(H, T, H, T) on four flips of a coin?

17. Solve for a. $6a+5=-31$

18. Multiply. $(x+8)(x-7)$

19. Evaluate. $5\cdot3+4\cdot3+10\div2$

20. Of the ten marbles in a bag, four are black, three are red, two are green, and one is white. What is the P(black, white) with replacement?

1.	2.	3.	4.
5.	6.	7.	8.
9.	10.	11.	12.
13.	14.	15.	16.
17.	18.	19.	20.

Lesson #118

1. Write the slope and the y-intercept of the equation $y = \frac{7}{6}x + 3$.

2. Find the solution for $4x + 2 < 10$.

3. In a survey of 3,000 people, 35% report having more than 4 televisions in their homes. How many of the people surveyed have more than 4 televisions?

4. Solve for h. $h - 19 = -50$

5. Find the median of 18, 35, 27, 56, 32 and 16.

6. Write 4,200 in scientific notation.

7. Evaluate the expression $5n + 3p - np$ if $n = 4$ and $p = 3$.

8. $0.007 \times 0.04 = ?$

9. How many centimeters are in 25 meters?

10. Simplify. $\dfrac{10x^4 + 40x^3}{2x^3 - 32x}$

11. Write the quadratic formula.

12. Multiply. $3c^2\left(4c^2 + 5c - 3\right)$

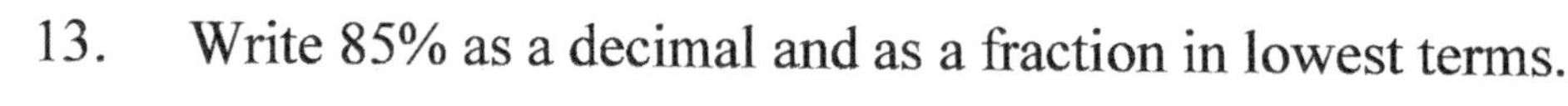

13. Write 85% as a decimal and as a fraction in lowest terms.

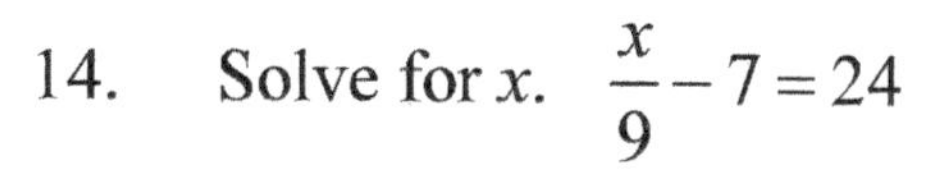

14. Solve for x. $\frac{x}{9} - 7 = 24$

15. Find the area of the trapezoid.

18 in.

7 in.

12 in.

16. Solve for x. $3x + 2x + 5 = 15$

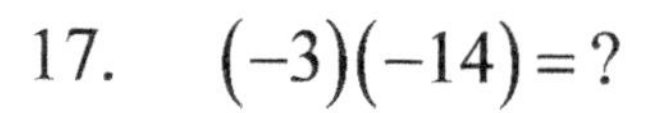

17. $(-3)(-14) = ?$

18. Simplify. $-\sqrt{16y^{16}}$

19. Multiply. $\left(4x^2 + 5x - 3\right)(x + 2)$

20. Find the sum.

$$\begin{array}{r} 15a^3 - 8a^2 + 3a - 4 \\ +\ 8a^3 + 2a^2 \quad\quad + 7 \\ \hline \end{array}$$

1.	2.	3.	4.
5.	6.	7.	8.
9.	10.	11.	12.
13.	14.	15.	16.
17.	18.	19.	20.

Lesson #119

1. Multiply. $(3x+5)(2x-4)$

2. $133+(-75)=?$

3. Simplify. $6^{-3}a^4b^{-2}$

4. Factor. $3y^2+2y-5$

5. Evaluate. $54 \div 6 + 3 \cdot 5 + 12 \div 3$

6. Solve for b. $b+23=92$

7. Simplify. $\dfrac{a-6}{a^2-6a}$

8. Write the slope-intercept form of a linear equation.

9. Use the quadratic formula to find the solution. $a^2+6a+8=0$

10. Simplify. $(4x^2y^4)^3$

11. Solve for x. $\dfrac{5}{x}=\dfrac{75}{105}$

12. Simplify. $\dfrac{a^{-4}}{a^{-8}}$

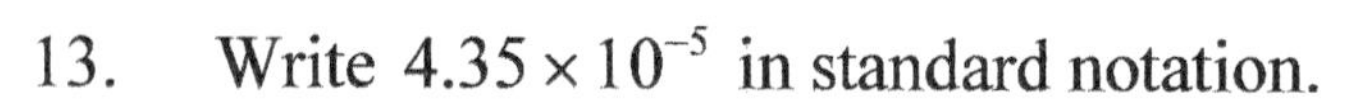

13. Write 4.35×10^{-5} in standard notation.

14. $\dfrac{6}{10} \div \dfrac{2}{5} = ?$

15. Solve for x. $4x+2x-6=42$

16. $0.3-0.1936=?$

17. What is the area of a triangle if the base is 16 m long and it is 5 m high?

18. Jake bought a new snow board for \$175. The snow board was on sale for 25% off, and the sales tax was 8%. How much did Jake pay for the snow board with the discount and sales tax?

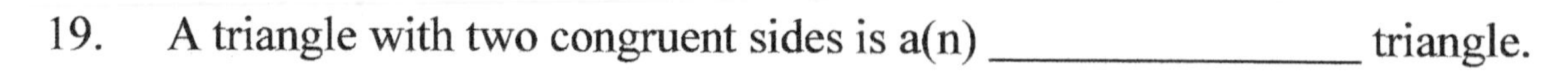

19. A triangle with two congruent sides is a(n) _______________ triangle.

20. Simplify. $-\sqrt{81a^{12}b^{18}}$

1.	2.	3.	4.
5.	6.	7.	8.
9.	10.	11.	12.
13.	14.	15.	16.
17.	18.	19.	20.

Lesson #120

1. Multiply. $(x+9)(x-3)$

2. Solve for x. $8x-12=5x$

3. $5.624 \div 0.04 = ?$

4. Simplify. $4n^3 \cdot 3n^5$

5. Write 0.000072 in scientific notation.

6. Solve for c. $\frac{1}{3}c=2$

7. Factor this expression. $3a^2-9a+15$

8. Solve for x. $\frac{7}{9}x-\frac{6}{9}x-4=12$

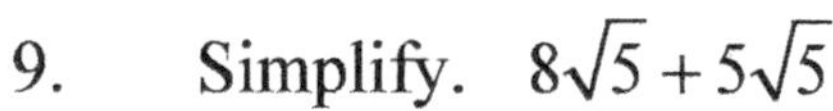

9. Simplify. $8\sqrt{5}+5\sqrt{5}$

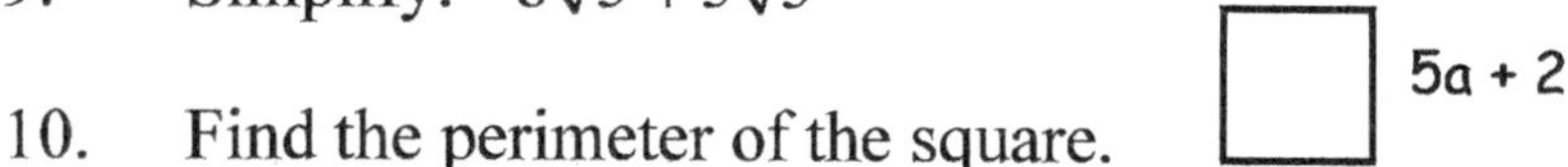

10. Find the perimeter of the square.

11. Rewrite as an algebraic expression: *Nine times a number decreased by four.*

12. $-29+(-18)+(-32)=?$

13. Simplify. $\frac{2x-10}{x-5}$

14. Two angles whose measures add up to 90° are _______________ angles.

15. Find the values for y in the equation $y=-2x+2$ when $x = \{0, -3, 1\}$.

16. $\frac{7}{15}\cdot\frac{12}{21}=?$

17. Simplify. $(-3)^{-2}$

18. Simplify. $(g^{10})^5$

19. Write $\frac{3}{25}$ as a decimal and as a percent.

20. Solve the system of equations using any method you choose. $x+y=12$, $x-y=2$

1.	2.	3.	4.
5.	6.	7.	8.
9.	10.	11.	12.
13.	14.	15.	16.
17.	18.	19.	20.

Lesson #121

1. $\begin{pmatrix} 5 & -7 \\ 2 & 0 \end{pmatrix} - \begin{pmatrix} -3 & 4 \\ 6 & -5 \end{pmatrix} = ?$

2. Multiply. $(4x-2)(x-6)$

3. Solve for x. $-3x=42$

4. Calculate the area of the parallelogram.

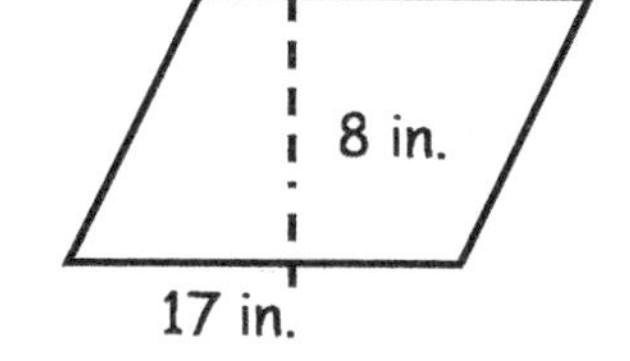

5. Simplify. $8x^{-2}y^{2}z^{-5}$

6. $-63+(-52)=?$

7. What is the percent of change from 4.2 to 6.4? Round your answer.

8. Find the value of x. $\frac{4}{7}=\frac{x}{105}$

9. $13\frac{3}{5}+9\frac{1}{4}=?$

10. Simplify. $(3x)^0$

11. The slope of a vertical line is __________.

12. Solve for t. $t-29=-63$

13. Simplify. $\sqrt{27}$

14. Multiply. $5n^2(4n+3)$

15. Solve for a. $\frac{4}{5}a=20$

16. Simplify. $\frac{18-3x}{x^2-36}$

17. Factor. $m^2-10m+25$

18. Which is greater, $\frac{3}{20}$ or 20%?

19. Find the LCM of $10a^3b^2c^5$ and $12a^2bc^3$.

20. Find the difference. $(12x^3-8x^2+9)-(16x^3-3)$

1.	2.	3.	4.
5.	6.	7.	8.
9.	10.	11.	12.
13.	14.	15.	16.
17.	18.	19.	20.

Lesson #122

1. Simplify. $3^{-2}a^2b^{-4}c$

2. $66+(-19)=?$

3. Write the formula for finding the area of a trapezoid.

4. Simplify. $4\sqrt{7}+5\sqrt{7}$

5. Evaluate. $32 \div 4 + 20 \div 5 + 3 \cdot 2$

6. Write the slope and the y-intercept for the line. $y=\frac{3}{4}x-\frac{1}{2}$

7. Multiply. $(5a+2)(6a-2)$

8. Write 0.0000082 in scientific notation.

9. Factor out the GCF. $16x^2-8x+4$

10. Write 80% as a decimal and as a reduced fraction.

11. Simplify. $\sqrt{36a^{10}b^{16}}$

12. Write the quadratic formula.

13. Solve for p. $p-27=72$

14. What is the percent of change from 15 miles to 12 miles?

15. Solve for a. $4a+2a-6=30$

16. $17\frac{1}{8}-8\frac{5}{8}=?$

17. Find the perimeter of a heptagon if its sides each measure 6 meters.

18. Evaluate the expression $3ab+2b$ if $a=2$ and $b=4$.

19. Solve for x. $3x+5=5x-5$

20. Simplify. $\frac{(b+2)(b-2)}{(b-2)(b-2)}$

1.	2.	3.	4.
5.	6.	7.	8.
9.	10.	11.	12.
13.	14.	15.	16.
17.	18.	19.	20.

Lesson #123

1. Solve for x. $\frac{3}{7}=\frac{x}{42}$

2. Lenny has a rope that is $22\frac{1}{2}$ inches long. He needs to cut the rope into five equal sections for his school project. How long will each piece of rope be?

3. Which is greater, $\frac{8}{25}$ or 36%?

4. $-25+(-13)+(-33)=?$

5. Write 4.7×10^4 in standard notation.

6. Multiply. $(7a^2-4a+2)(a+5)$

7. Simplify. $(7a^3b^2c^4)^2$

8. $42\frac{1}{7}+36\frac{2}{5}=?$

9. $-150-76=?$

10. Find the area of the square.

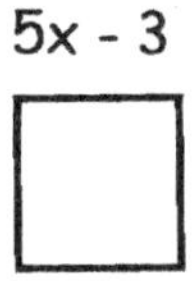

11. Evaluate. $62+14\div 2+3\cdot 5$

12. Solve for t. $5(t+1)=10$

13. Factor the polynomial. $y^2-9y+14$

14. Simplify. $\frac{7ab^{-2}}{3c^{-3}}$

15. Solve for x. $4x-9\geq 15$

16. Simplify. $\sqrt{64b^6c^8}$

17. Simplify. $4\sqrt{3}-\sqrt{12}$

18. Write an algebraic phrase for *the quotient of a number and nine decreased by four.*

19. Find the quotient. $\frac{x}{x+4}\div\frac{x+3}{x+4}$

20. Find the difference.

$$\begin{array}{r} 10a^3-6a^2+4 \\ -\;\;6a^3+2a^2-4a-3 \\ \hline \end{array}$$

1.	2.	3.	4.
5.	6.	7.	8.
9.	10.	11.	12.
13.	14.	15.	16.
17.	18.	19.	20.

Lesson #124

1. $-55+(-63)=?$

2. Simplify. $\left(8x^2y^4\right)^3$

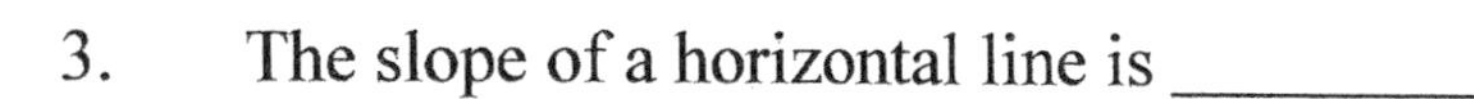

3. The slope of a horizontal line is ________.

4. Factor out the GCF. $10x^2-5x+5$

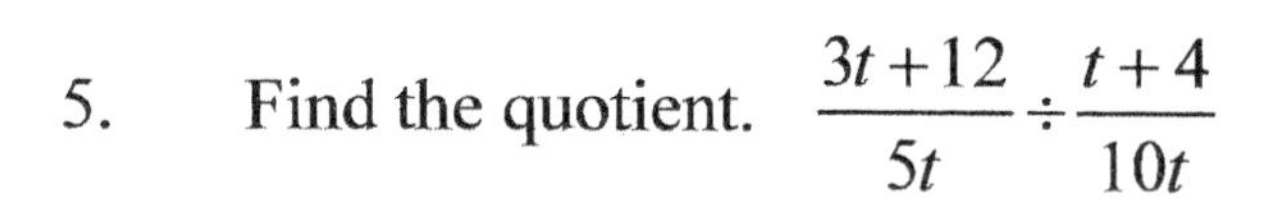

5. Find the quotient. $\dfrac{3t+12}{5t}\div\dfrac{t+4}{10t}$

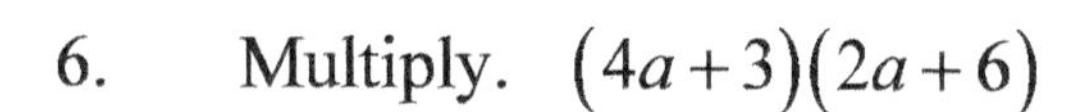

6. Multiply. $(4a+3)(2a+6)$

7. Write the formula for finding the circumference of a circle.

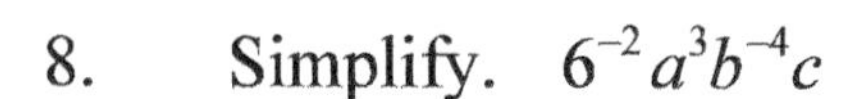

8. Simplify. $6^{-2}a^3b^{-4}c$

9. Find the missing measurement, x.

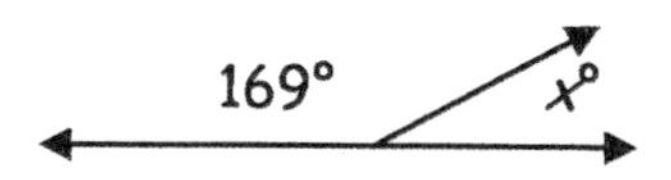

10. A triangle with all sides congruent is called a(n) ____________ triangle.

11. How many feet are in 8 miles?

12. Factor. $d^2-7d+12$

13. Find the surface area.

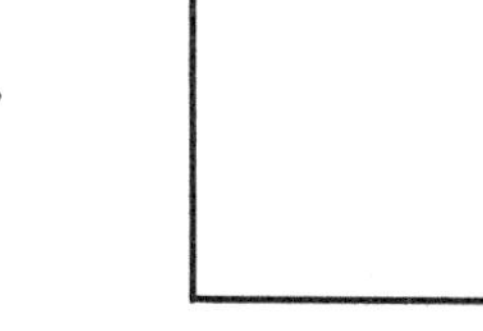

14. Simplify. $\sqrt{\dfrac{4}{25}}$

15. What is the P(H, T, H, T, H, T) on six flips of a coin?

16. Water boils at _______ °F.

17. $32.5\times0.3=?$

18. Write 0.000413 in scientific notation.

19. $\sqrt{7921}=?$

20. The perimeter of the triangle is (23a – 7). Find the length of the third side.

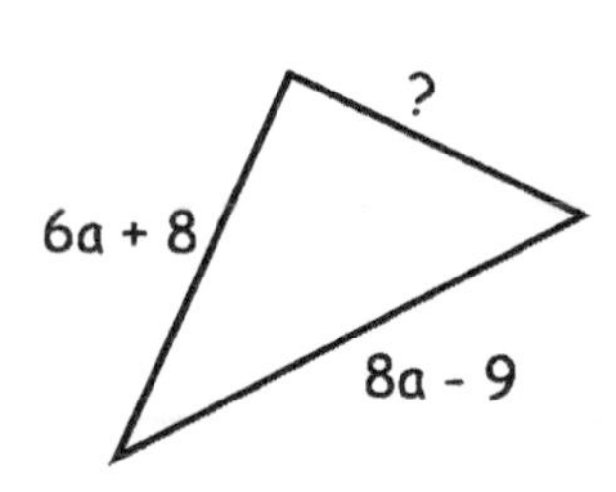

1.	2.	3.	4.
5.	6.	7.	8.
9.	10.	11.	12.
13.	14.	15.	16.
17.	18.	19.	20.

Lesson #125

1. Simplify. $\dfrac{24y+18}{36}$

2. $72-(-48)=?$

3. Find the area of the square.

4. Simplify. $3^{-3}a^2b^{-4}c$

5. Multiply the rational expression. $\dfrac{6x^4}{5y^7}\cdot\dfrac{3y^5}{8x}$

6. Write the slope-intercept form of a linear equation.

7. Multiply. $(3x-4)(x+2)$

8. $0.63-0.3875=?$

9. Write 4.16×10^5 in standard notation.

10. Factor. $9y^2-25$

11. Which is greater, $\dfrac{3}{50}$ or 10%?

12. Factor out the GCF. $6b^3-18b^2+12$

13. Simplify. $\left(4c^2d^3f^4\right)^2$

14. Solve for x. $x+9=2x-6$

15. Solve for p. $p-39=86$

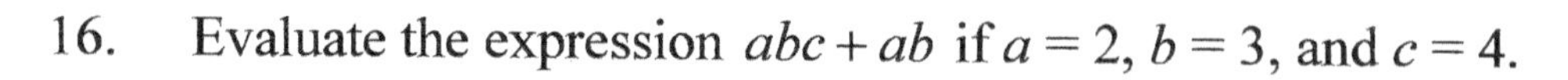

16. Evaluate the expression $abc+ab$ if $a=2$, $b=3$, and $c=4$.

17. Simplify. $3x\left(2x^2\right)\left(5x^3\right)$

18. Tanya filled her backpack with 9 books. Each book weighs 20 ounces. The empty backpack weighs 3 pounds. How much does the full backpack weigh?

19. Simplify. $10\sqrt{3}-6\sqrt{3}$

20. Solve for x. $\dfrac{x}{7}+7=20$

1.	2.	3.	4.
5.	6.	7.	8.
9.	10.	11.	12.
13.	14.	15.	16.
17.	18.	19.	20.

Lesson #126

1. Solve for x. $5x-7=18$

2. Simplify. $-\sqrt{100x^8y^{12}}$

3. Solve the system using any method you choose. $\begin{aligned} x-y&=12 \\ x+y&=22 \end{aligned}$

4. Write 0.0000012 in scientific notation.

5. Find the slope of a line passing through points (0, 4) and (2, 2).

6. $132-(-59)=?$

7. Simplify. $\sqrt{50}$

8. Multiply. $\dfrac{x^2+13x+42}{x^2-3x-40}\cdot\dfrac{x-8}{x+6}$

9. Factor. $a^2-9a+14$

10. Simplify. $2\sqrt{4y^3}$

11. Multiply. $5d\left(3d^2-4d\right)$

12. Divide. $\dfrac{m^2}{p^2}\div\dfrac{p^4}{m^4}$

13. Simplify. $\sqrt{\dfrac{25}{49}}$

14. Find the volume of the rectangular prism.

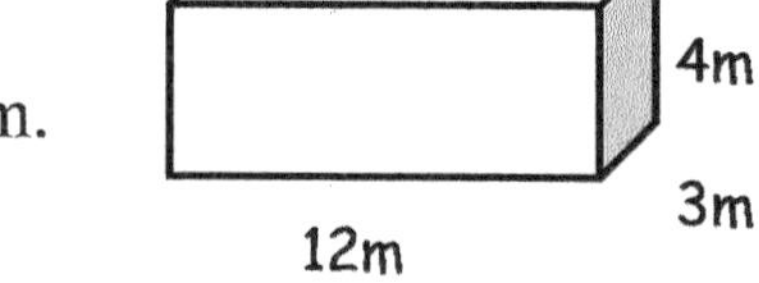

15. Simplify. $a\cdot a^{-7}$

16. 80% of what number is about 25? Round your answer to the nearest whole number.

17. Write the formula for finding the area of a circle.

18. Multiply. $(x+9)(x-3)$

19. Solve for m. $-5m=-70$

20. Give the coordinates of points B and D.

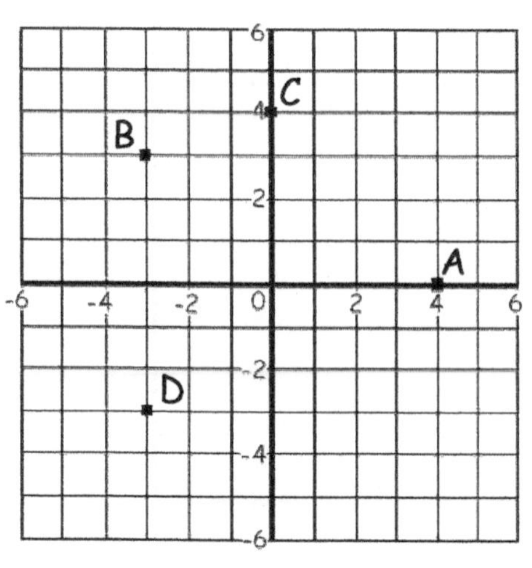

1.	2.	3.	4.
5.	6.	7.	8.
9.	10.	11.	12.
13.	14.	15.	16.
17.	18.	19.	20.

Lesson #127

1. Factor. $y^2 - 16y + 64$

2. $-3(-8)(2) = ?$

3. Multiply. $(5a^3 - 2a^2 + 4)(a - 6)$

4. Solve for x. $\frac{3}{8} = \frac{x}{120}$

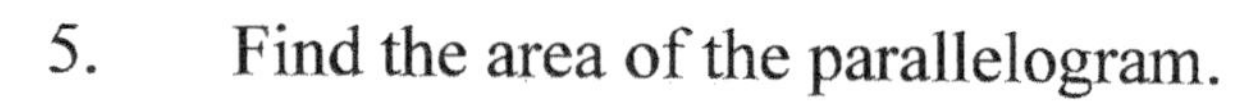

5. Find the area of the parallelogram.

4y

$y^2 + 2$

6. $239 + (-99) = ?$

7. Simplify. $4\sqrt{18a^{17}b^{15}}$

8. $50 - 4 \cdot 5 + 6 \div 2 - 1 = ?$

9. Solve for x. $\frac{8}{9}x - \frac{7}{9}x + 4 = 12$

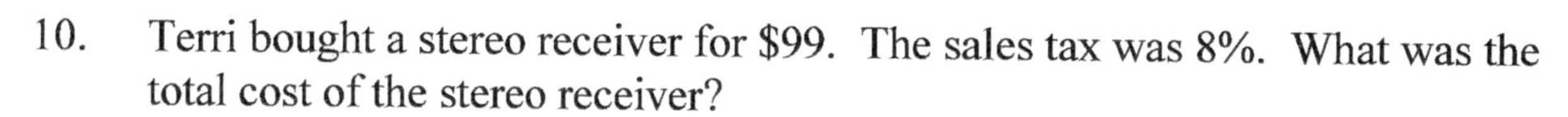

10. Terri bought a stereo receiver for \$99. The sales tax was 8%. What was the total cost of the stereo receiver?

11. Multiply the rational expression. $\frac{x^2 + 7x + 12}{x + 5} \cdot \frac{5x + 25}{x + 4}$

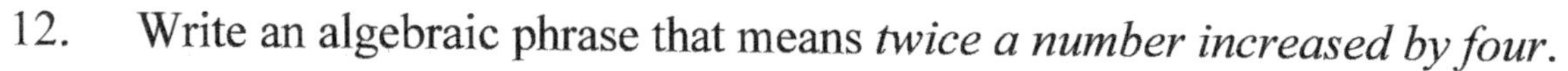

12. Write an algebraic phrase that means *twice a number increased by four.*

13. Find the LCM of $10x^3y^2z^4$ and $18x^2yz^2$.

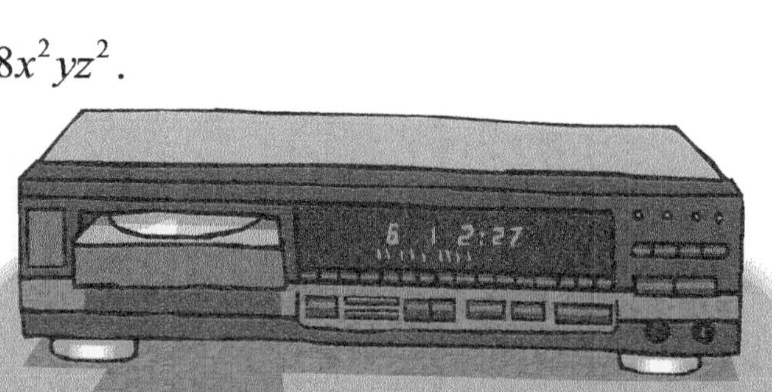

14. 70% of what number is 63?

15. Factor out the GCF. $5x - 20$

16. Solve for y. $|y - 3| = 7$

17. Solve for m. $14 - 9m = -11m$

18. Simplify. $(x^{-8}y^{-5})(x^{10}y^{-2})$

19. Write 52,000,000 in scientific notation.

20. Find the sum.
$$\begin{array}{r} 15x^2 - 7x + 3 \\ +\ 8x^2 \quad\ \ - 5 \\ \hline \end{array}$$

1.	2.	3.	4.
5.	6.	7.	8.
9.	10.	11.	12.
13.	14.	15.	16.
17.	18.	19.	20.

Lesson #128

1. Simplify. $\left(d^4\right)^{-7}$

2. Multiply. $(6x-3)(x+5)$

3. Solve for x. $\frac{x}{12}+4=15$

4. Factor. $y^2+9y+14$

5. Find $\frac{7}{8}$ of 120.

6. Simplify. $\frac{3x^3-12x}{6x^4-12x^3}$

7. $-215+(-138)=?$

8. Solve for c. $c+10=-31$

9. Evaluate $\frac{ab}{4}+2b$ if $a=2$ and $b=8$.

10. Use the quadratic equation to solve. $(x+4)(3x-15)=0$

11. Simplify. $a^{-4}b^2cd^{-3}$

12. Find the perimeter of the triangle.

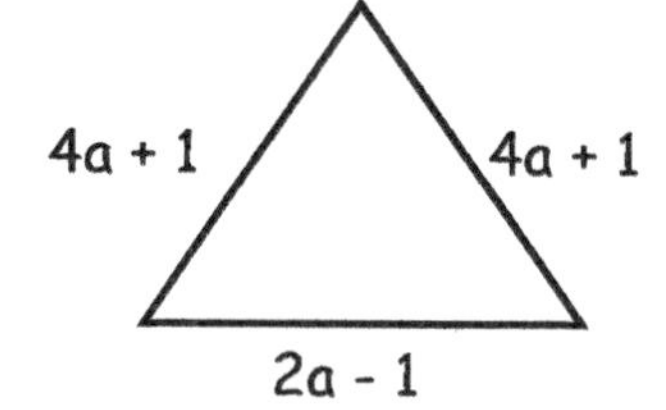

13. Solve for b. $-7b=98$

14. Solve for x. $4x+2x-6=30$

15. Write an algebraic phrase for *the sum of a number and ten.*

16. Write 3.13×10^{-4} in standard form.

17. Find the percent of change from 14 miles to 24 miles. Round your answer to the nearest whole number.

18. $3a-4>8$ Graph the solution on a number line.

19. Write $\frac{9}{50}$ as a decimal and a percent.

20. The ratio of roses to tulips in the garden is 4 to 7. If there are 64 roses, how many tulips are in the garden?

1.	2.	3.	4.
5.	6.	7.	8.
9.	10.	11.	12.
13.	14.	15.	16.
17.	18.	19.	20.

Lesson #129

1. Find the slope of a line passing through points (0, 5) and (4, 9).

2. Solve for x. $\frac{x}{7}=14$

3. Write an algebraic phrase to represent *the quotient of a number and six decreased by eight.*

4. $\frac{-224}{-4}=?$

5. Simplify. $\frac{(5x-3)(x+9)}{(x+1)(5x-3)}$

6. $|-36|=?$

7. Factor. c^2-c-56

8. Multiply. $\frac{6}{10x}\cdot\frac{5}{2}$

9. Multiply. $(8x^2+6x-3)(3x-2)$

10. Find the area of the triangle.

7 in.

12 in.

11. $1.6+38.24+9.573=?$

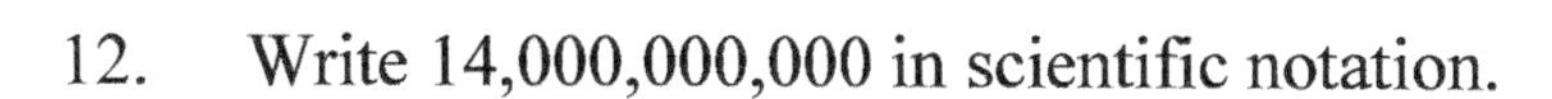

12. Write 14,000,000,000 in scientific notation.

13. Water boils at ________°C.

14. Factor out the GCF. $24a^3+12a^2-6a$

15. Solve for a. $25-5a=6a+3$

16. Put these in increasing order. 0.45 $\frac{2}{5}$ 48%

17. Simplify. $\sqrt{45m^7n^5}$

18. $-18 \bigcirc -55$

19. $18\frac{2}{5}+12\frac{2}{3}=?$

20. Simplify. $\frac{\sqrt{50x^3}}{\sqrt{2x}}$

1.	2.	3.	4.
5.	6.	7.	8.
9.	10.	11.	12.
13.	14.	15.	16.
17.	18.	19.	20.

Lesson #130

1. $\frac{8}{12} \div \frac{2}{3} = ?$

2. Simplify. $5\sqrt{24x^{13}}$

3. Factor. $4a^2 - 11a - 3$

4. $0.009 \times 0.006 = ?$

5. Simplify. $\frac{8a^3b^2c}{12abc}$

6. Multiply. $(x+9)(x+5)$

7. Simplify. $(3a^3)(6a^2)(2a)$

8. $\sqrt{196} = ?$

9. Find the perimeter of the polygon.

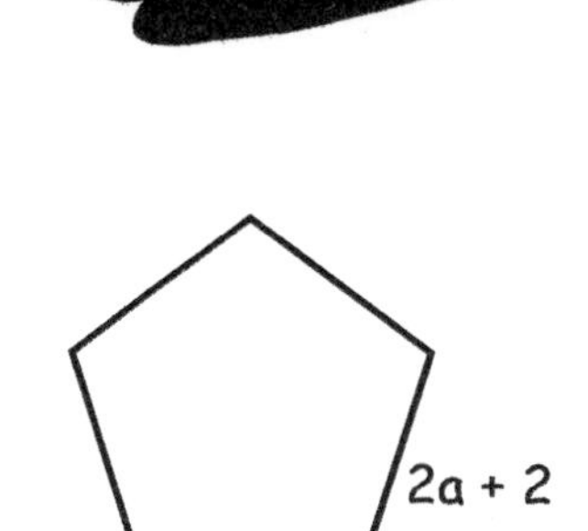

10. $-4(9)(-2) = ?$

11. $-2y \leq 16$ Find the solution to the inequality and graph it.

12. Simplify. $(x^3y^4z^2)^4$

13. A roll of ribbon contains 12 feet of ribbon. Mrs. Benson used 32 inches of ribbon to wrap a package. How much ribbon was left on the roll?

14. Simplify. $4^0x^2y^{-3}$

15. What is the P(H, H, H, T, T, T,) on six flips of a coin?

16. Solve for y. $|y-6| = 10$

17. Find the value of x. $\frac{3}{5}x = 45$

18. Find the circumference of a circle if the radius is 7 inches.

19. Solve for x. $\frac{2}{9} = \frac{x}{108}$

20. Multiply. $\frac{6x^2}{2x-4} \cdot \frac{x-2}{1}$

1.	2.	3.	4.
5.	6.	7.	8.
9.	10.	11.	12.
13.	14.	15.	16.
17.	18.	19.	20.

Lesson #131

1. $-4 < 7 + x < 2$ Graph the solution to the inequality.

2. $162 + (-88) = ?$

3. Solve for y. $|y - 4| < 7$

4. Multiply. $(4x + 7)(2x - 3)$

5. Write 5.16×10^7 in standard notation.

6. Find the difference. $(15x^3 - 8x^2 + 4x - 9) - (8x^3 + 3x^2 - 2x + 4)$

7. Multiply. $4d(3d^3 - 5d^2 + 6)$

8. Factor. $s^2 - 11s + 28$

9. Simplify. $4x^4y^{-5}z^{-2}$

10. Determine the area of the square.

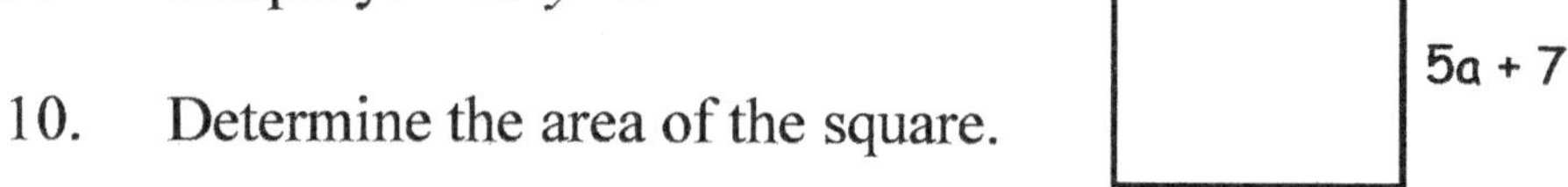

11. Simplify. $9\sqrt{36x^{32}}$

12. Write the slope-intercept form of a linear equation.

13. A triangle in which all three sides have different lengths is ___________.

14. $0.56 - 0.2134 = ?$

15. Solve for x. $x - 14 = -5$

16. Evaluate the expression $rst - s$ if $r = 3$, $s = 2$, and $t = 4$.

17. Find $\frac{5}{9}$ of 72.

18. Solve for x. $\frac{x}{8} - 5 = 10$

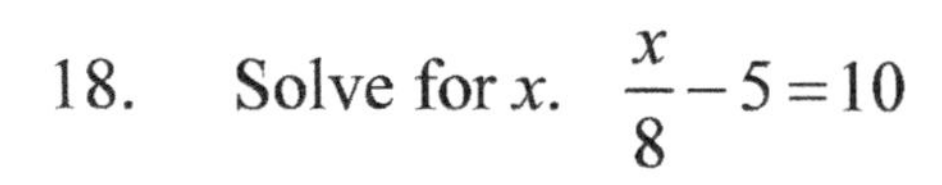

19. What is the average of 65, 75 and 85?

20. Simplify. $\frac{10}{3b} - \frac{8}{3b}$

1.	2.	3.	4.
5.	6.	7.	8.
9.	10.	11.	12.
13.	14.	15.	16.
17.	18.	19.	20.

Lesson #132

1. Factor. $c^2 - 36$

2. Write 82,000 in scientific notation.

3. Solve for y. $-3y = 39$

4. Find the area of the trapezoid.

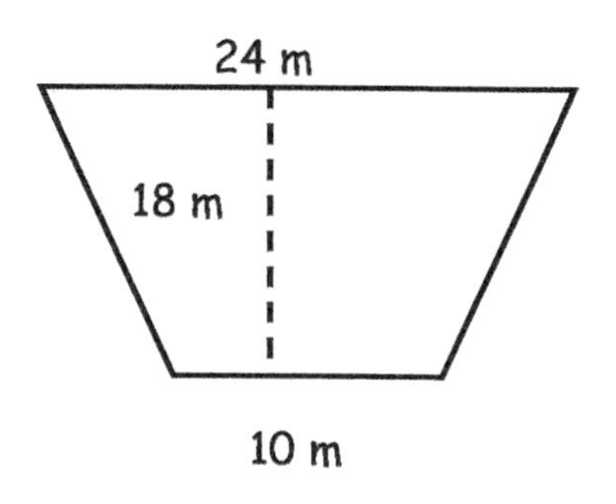

5. Write an algebraic phrase that means *five less than three times a number.*

6. Simplify. $3^{-3}x^4z^5$

7. How many grams are in 18 kilograms?

8. $-34 + (-17) + 11 = ?$

9. Multiply. $(6a^2 + 2a + 4)(a + 3)$

10. Simplify. $(g^5)^5$

11. $32 \div 4 + 3 \cdot 5 - 2 \cdot 1 = ?$

12. Find the percent of change from 8 to 12 feet.

13. Factor out the GCF. $8b^3 - 16b^2 + 4b$

14. Solve for h. $h + 31 = -85$

15. Find the values for y in the equation $y = 2x + 1$ when $x = \{0, -3, 4\}$.

16. Give the formula for finding the volume of a cube.

17. Simplify. $2\sqrt{36c^5}$

18. Write $\frac{7}{50}$ as a decimal and a percent.

19. Solve for x. $4x + 2x - 6 = 42$

20. Multiply. $\frac{12(a-2)}{a-3} \cdot \frac{a-3}{6(a+5)}$

1.	2.	3.	4.
5.	6.	7.	8.
9.	10.	11.	12.
13.	14.	15.	16.
17.	18.	19.	20.

Lesson #133

1. Solve for x. $5x-6=2x$
2. Simplify. $3ab^3\left(a^2b^4\right)$
3. $-99+(-47)=?$
4. Find the product. $\left(5x^2-3x+2\right)(x-5)$
5. Write the slope and the y-intercept of the line whose equation is $y=7x-3$.
6. $\dfrac{-64}{-4}=?$
7. Factor. $x^2-16x+48$
8. Simplify. $9\sqrt{27a^{15}b^9}$
9. What is the formula for finding the area of a circle?
10. Write 2.5×10^{-6} in standard notation.
11. Multiply. $6a\left(4a^2+7a-9\right)$
12. Add. $\dfrac{7x}{3}+\dfrac{2x}{3}$
13. $\begin{pmatrix}6 & -3 & 4\\5 & -2 & 1\end{pmatrix}+\begin{pmatrix}5 & -2 & -3\\0 & 5 & -6\end{pmatrix}=?$
14. 80% of 35 is what number?
15. $25+3[4+2(3)-2]=?$
16. Solve for x. $\dfrac{x}{9}=12$
17. Solve for x. $2x+3(x-2)=24$
18. Divide. $\dfrac{x^2}{3y}\div\dfrac{6x}{4y^3}$
19. What is the sum? $\dfrac{5b}{21}+\dfrac{2b}{21}+\dfrac{8b}{21}$
20. Find the perimeter of the square.

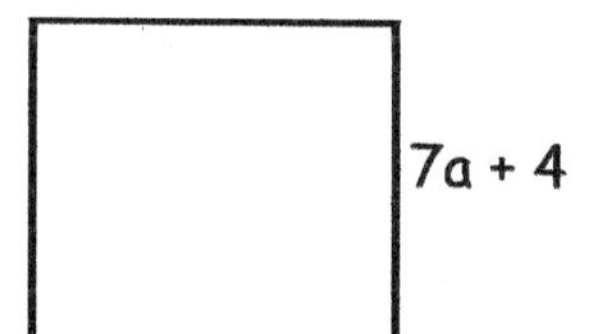

1.	2.	3.	4.
5.	6.	7.	8.
9.	10.	11.	12.
13.	14.	15.	16.
17.	18.	19.	20.

Lesson #134

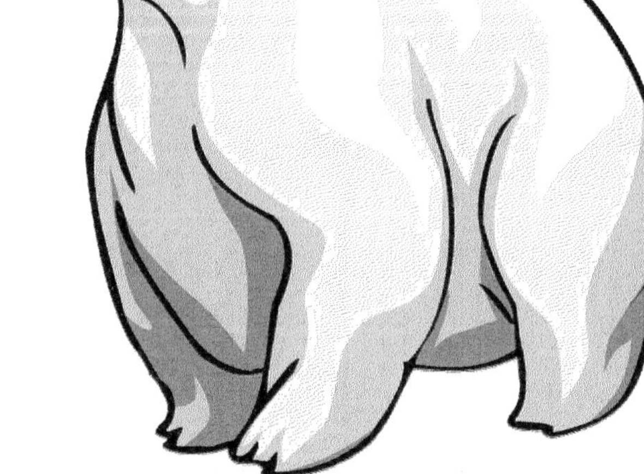

1. Solve for x. $7x-4=24$
2. Simplify. $\left(h^4\right)^{-3}$
3. $-66+(-29)=?$
4. Evaluate $5a-2b+ab$ if $a=3$ and $b=2$.
5. Write the equation of a line through $(3,-1)$ with a slope of 1.
6. Solve for x. $\frac{3}{2}=\frac{18}{x}$
7. What value of h will make this true? $4h+5=9h$
8. Write 910,000,000,000 in scientific notation.
9. Factor. t^2-6t+9
10. Solve using any method you choose. $x+y=19$, $x-y=-7$
11. A jar has five blue balls, three yellow, six green, and two red balls. If you pick two balls from the jar, what is the P(R, B) with replacement? What is the P(Y, G) without replacement?
12. Simplify. $\left(6x^3\right)(3x)\left(2x^2\right)$
13. Find the percent of change from 18 kg to 20 kg. Round your answer.
14. What is the solution to this inequality? $37<3c+7<43$
15. Multiply. $8m(4m-5)$
16. Factor out the GCF. $5r+20r^3+15r^2$
17. Write the quadratic formula.
18. Multiply. $(3a-1)(2a+1)$
19. Find the perimeter of the rectangle.

20. Simplify.

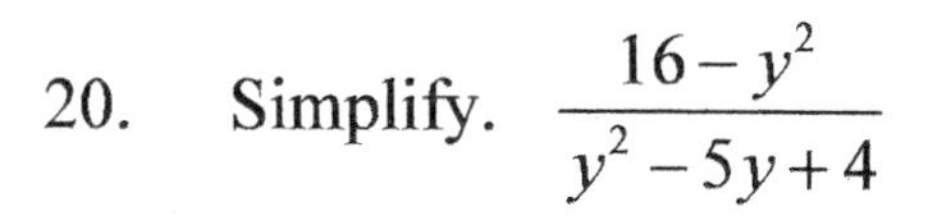

$\frac{16-y^2}{y^2-5y+4}$

1.	2.	3.	4.
5.	6.	7.	8.
9.	10.	11.	12.
13.	14.	15.	16.
17.	18.	19.	20.

Lesson #135

1. $-12(-3)(-4)=?$

2. Find the slope of a line passing through points (1, 3) and (5, 6).

3. Multiply. $(9s^2+3s-4)(2s+2)$

4. Factor. $x^2+7x+12$

5. Solve for x. $\frac{x}{10}-4=6$

6. Write 0.00000913 in scientific notation.

7. Simplify. $(3x^3)(5x^2)(2x)$

8. Multiply. $7c^2(4c^2+3c-5)$

9. How many cups are in 36 pints?

10. Solve for x. $\frac{4}{7}x=-28$

11. Find the area of the triangle.

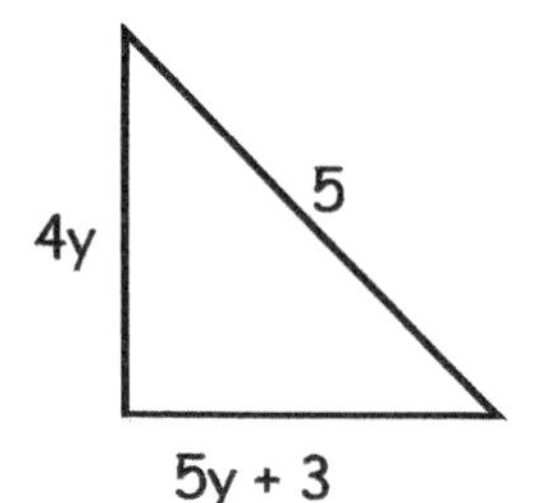

12. Simplify. $(9a^5b^3c^2)^2$

13. Write $\frac{9}{20}$ as a decimal and as a percent.

14. Solve for x. $5x+3=2x+6$

15. Simplify. $5\sqrt{49x^{12}y^8}$

16. Find the LCM of $14a^3b^2$ and $16a^2bc^2$.

17. Multiply. $\frac{9y^2}{4-2y}\cdot\frac{5y-10}{21y}$

18. Simplify. $4\sqrt{7}+9\sqrt{7}$

19. $56\frac{2}{9}+43\frac{1}{5}=?$

20. Simplify. $\frac{4x}{x-3}-\frac{3x}{x-3}$

1.	2.	3.	4.
5.	6.	7.	8.
9.	10.	11.	12.
13.	14.	15.	16.
17.	18.	19.	20.

Lesson #136

1. Put these integers in increasing order. –18, –36, 0, –2, –25

2. Solve for h. $|h-4|=9$

3. $2x+3\leq 9$ Graph the solution on a number line.

4. $136+(-98)=?$

5. Simplify. $\dfrac{3p-18}{2p-12}$

6. $15 \div 3 + 4 \cdot 5 - 12 \div 2 = ?$

7. Write *five less than a number* using an algebraic phrase.

8. $17 - 9\frac{3}{7} = ?$

9. Write 2.8×10^9 in standard notation.

10. Solve for c. $7c-9=8c$

11. Factor. a^2-49

12. Multiply. $(4x+3)(2x-1)$

13. Simplify. $-7\sqrt{25x^9}$

14. $4.53 \times 0.6 = ?$

15. Find the sum of $(10a^2-9)$ and $(5a^2+3)$.

16. Simplify. $\dfrac{12c^2d}{18cd^3}$

17. Solve using any system. $\begin{aligned} 3x+7y&=3 \\ x-7y&=1 \end{aligned}$

18. $\dfrac{8}{9}\times\dfrac{18}{32}=?$

19. Find the volume of the cube.

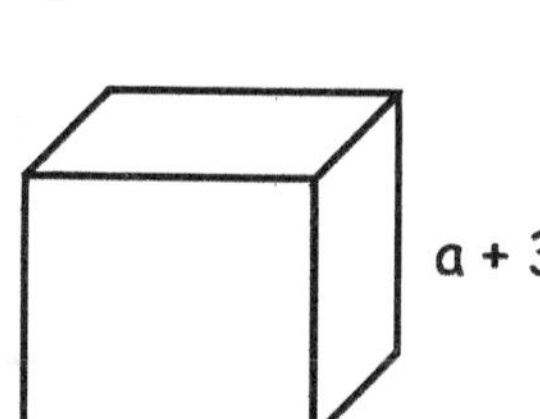

20. Subtract. $\dfrac{5a+4}{9a}-\dfrac{3a-1}{12a}$

1.	2.	3.	4.
5.	6.	7.	8.
9.	10.	11.	12.
13.	14.	15.	16.
17.	18.	19.	20.

Lesson #137

1. Solve for x. $3x-9=18$

2. Multiply. $(x-7)(x+6)$

3. Factor. $y^2-10y+16$

4. Simplify. $15\sqrt{9}-\sqrt{9}$

5. Solve for y. $2(y+4)=16$

6. Simplify. $\dfrac{(8x^2)(3x^4)}{2x}$

7. The slope of a horizontal line is ________.

8. Find the percent of change from 15 inches to 25 inches.

9. Simplify. $2^2c^{-2}d^3$

10. $-|-29|=?$

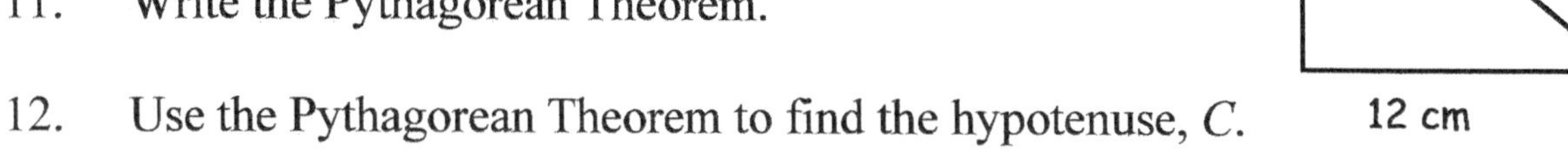

11. Write the Pythagorean Theorem.

12. Use the Pythagorean Theorem to find the hypotenuse, C.

13. Simplify. $\sqrt{x^9y^7}$

14. 36 is what percent of 80?

15. In a right triangle, what term represents the side opposite the right angle?

16. Write 371,000,000 in scientific notation.

17. Simplify. $\sqrt{27}$

18. Multiply. $9p(6p^2-4p+3)$

19. Find the difference. $(23x^3+15x^2+8)-(12x^3+10x^2-2)$

20. Add. $\dfrac{a-4}{3a-2}+\dfrac{a+3}{1}$

1.	2.	3.	4.
5.	6.	7.	8.
9.	10.	11.	12.
13.	14.	15.	16.
17.	18.	19.	20.

Lesson #138

1. Simplify. $\left(s^{10}\right)^{-4}$

2. Write the Pythagorean Theorem.

3. $16\frac{2}{7}-8\frac{5}{7}=?$

4. Solve for x. $\frac{3}{5}=\frac{x}{75}$

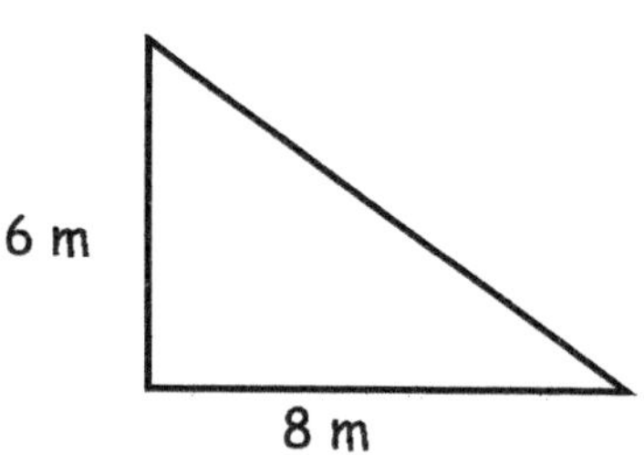

5. Find the length of the hypotenuse.

6. Write 2.9×10^{-3} in standard notation.

7. $47.2+9.365=?$

8. Simplify. $\frac{\left(4a^{3}\right)\left(3a^{2}\right)}{2a}$

9. Factor. $x^{2}-8x+16$

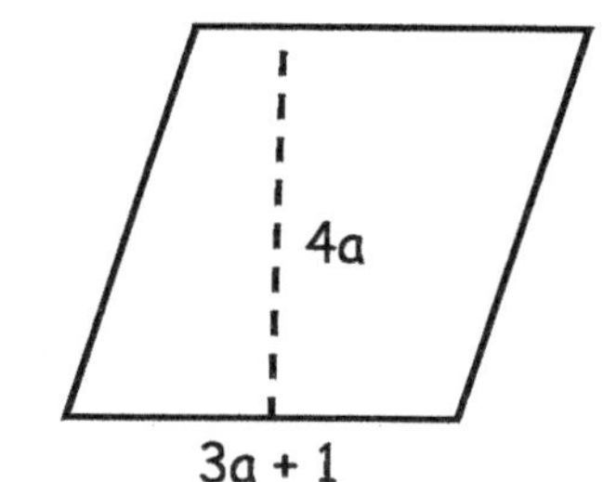

10. Which is greater, $\frac{7}{25}$ or 0.36?

11. Find the area of the parallelogram.

12. Simplify. $5\sqrt{16a^{5}b^{2}c^{7}}$

13. Multiply. $(5x+3)(2x-4)$

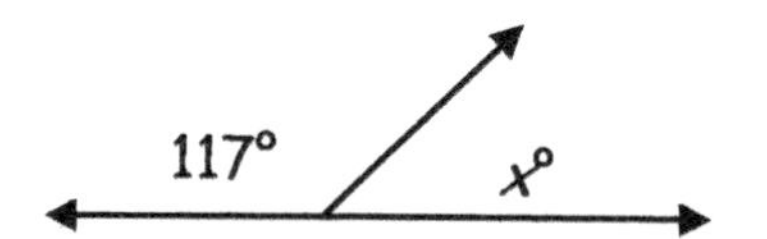

14. Find the missing measurement, x.

15. Write an equation of a line passing through point (4, 0) with a slope of 7.

16. Solve for a. $|a+6|=4$

17. Subtract. $\frac{a+2}{1}-\frac{2a}{a+1}$

18. Find the values for y in the equation $y=-4x+1$ when $x=\{-1, 2, -3\}$.

19. Solve for x. $\frac{x}{9}-7=8$

20. Simplify. $\frac{8a^{3}-4a^{4}}{5a^{3}-10a^{2}}$

1.	2.	3.	4.
5.	6.	7.	8.
9.	10.	11.	12.
13.	14.	15.	16.
17.	18.	19.	20.

Lesson #139

1. Find the coordinates of points A, C, and E.

2. 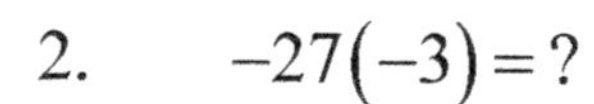
$-27(-3) = ?$

3. The slope of a vertical line is ___________.

4. Write the Pythagorean Theorem.

5. 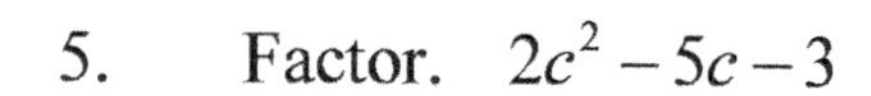
Factor. $2c^2 - 5c - 3$

6. Divide. $\dfrac{3x^3}{2} \div \dfrac{-15x^5}{1}$

7. Multiply. $(6b^2 - 4b + 2)(3b + 3)$

8. 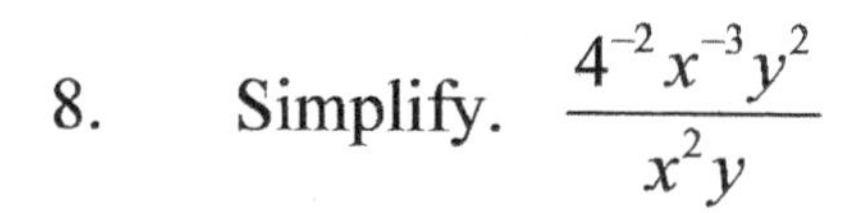
Simplify. $\dfrac{4^{-2}x^{-3}y^2}{x^2y}$

9. How many feet are in 5 miles?

10. 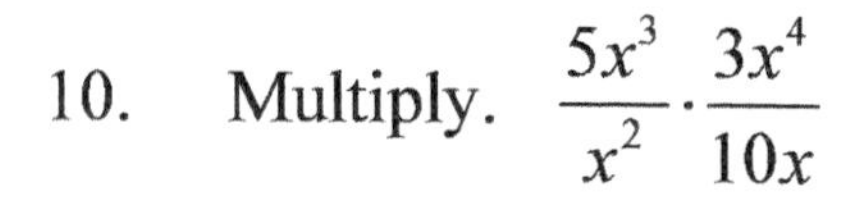
Multiply. $\dfrac{5x^3}{x^2} \cdot \dfrac{3x^4}{10x}$

11. Find the area of the rectangle.

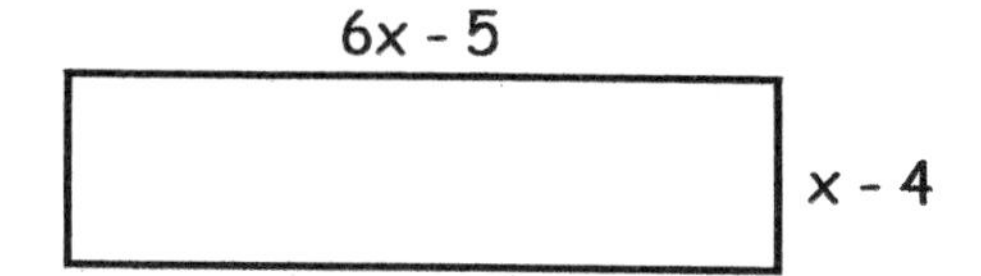

12. Simplify. $3\sqrt{18x^4y^8}$

13. Solve for x. $x - 27 = -56$

14. Subtract. $\dfrac{a}{2} - \dfrac{3a}{10}$

15. The perimeter of the rectangle to the right is $(38y + 2)$. Find its width.

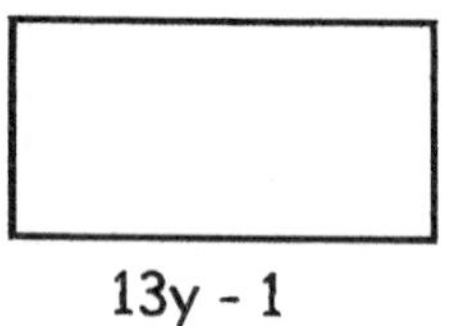

16. Write 0.0056 in scientific notation.

17. $19\frac{1}{8} + 24\frac{2}{5} = ?$

18. Solve for x. $7 - 3x > 4x - 21$

19. Find the length of the missing side.

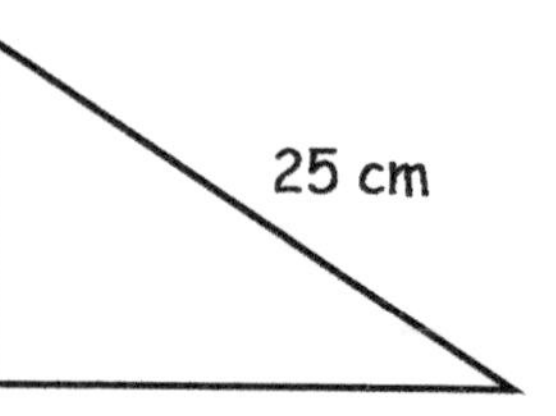

20. $-66 - (-42) = ?$

1.	2.	3.	4.
5.	6.	7.	8.
9.	10.	11.	12.
13.	14.	15.	16.
17.	18.	19.	20.

Lesson #140

1. $3x - 8 \geq 7$ Find the solution and graph it on a number line.

2. Simplify. $(5x^2y)(-2xy^2)$

3. Multiply. $(x+8)(x+5)$

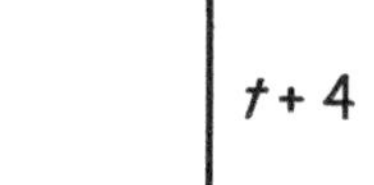

4. Find the area of the square.

5. Write the Pythagorean Theorem.

6. $-55 - (-24) = ?$

7. Multiply. $\frac{3x^7}{4y^5} \cdot \frac{2y^6}{3x^8}$

8. Write the slope-intercept form for a linear equation.

9. Simplify. $(2b^3c^2d^4)^5$

10. Simplify. $2\sqrt{20a^5b^{11}}$

11. Add. $\frac{3x-5}{4} + \frac{5x-3}{3}$

12. What is the P(1, 6, 3) on 3 rolls of a die?

13. Factor out the GCF. $8x^4 - 12x^3 + 16x^2$

14. Translate *sixteen times a number divided by twelve* as an algebraic phrase.

15. Evaluate $cd - c$ if $c = 10$ and $d = 4$.

16. $28 - 4[3 + 2(3 + 2) - 1] = ?$

17. Write 2.1×10^6 in standard notation.

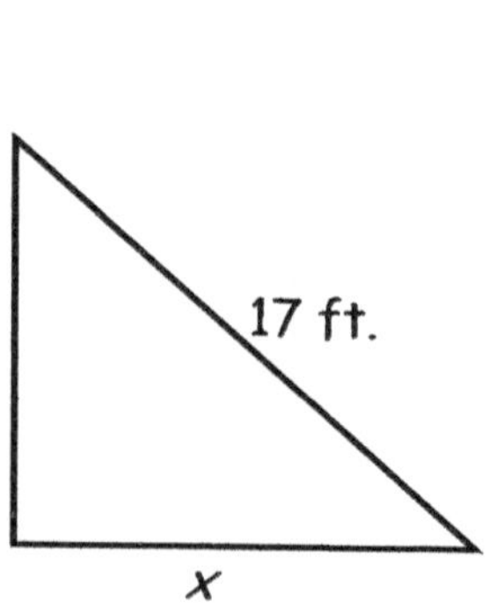

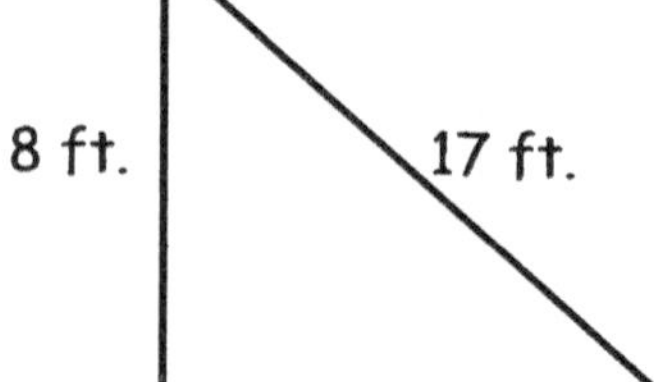

18. Find the missing length of the leg, x.

19. Factor. $x^2 - 15x + 56$

20. Find the difference.

$$\begin{array}{r} 0.4x^2 + 0.3x - 0.7 \\ -\ \ 0.2x^2 + 0.2x + 0.5 \\ \hline \end{array}$$

1.	2.	3.	4.
5.	6.	7.	8.
9.	10.	11.	12.
13.	14.	15.	16.
17.	18.	19.	20.

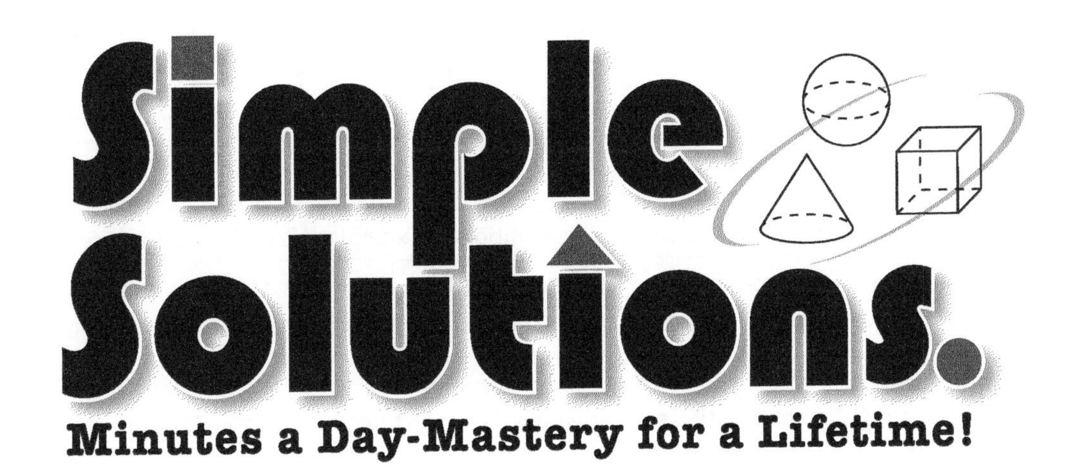

Algebra I

Part B

Help Pages

Help Pages

Vocabulary

General

Absolute Value — the distance between a number, x, and zero on a number line; written as $|x|$. Example: $|5| = 5$ reads "The absolute value of 5 is 5." $|-7| = 7$ reads "The absolute value of -7 is 7."

Binomial — a polynomial having exactly 2 terms. Examples: $3x - 7$, $2x + 5y$, $2y^2 + x^3$

Complementary Angles — two angles whose measures add up to 90°.

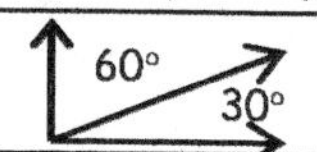

Expression — a mathematical phrase written in symbols. Example: $2x + 5$ is an expression.

Function — a rule that pairs each number in a given set (the domain) with just one number in another set (the range). Example: The function $y = x + 3$ pairs every number with another number that is larger by 3.

Hypotenuse — in a right triangle, the side opposite the right angle.

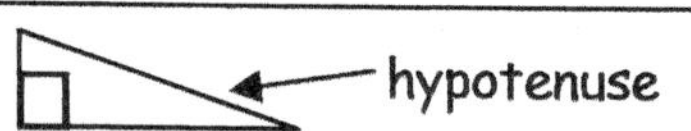

Integers — the set of whole numbers, positive or negative, and zero.

Irrational Number — a number that cannot be written as the ratio of two whole numbers. The decimal form of an irrational number is <u>neither</u> terminating nor repeating. Examples: $\sqrt{2}$ and π.

Legs — in a right triangle, the sides adjacent to the right angle. The two legs actually form the right angle.

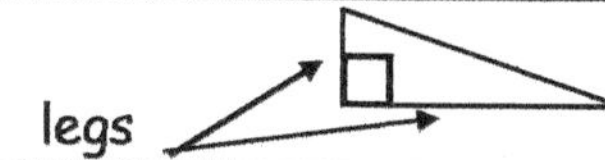

Matrix — a rectangular arrangement of numbers in rows and columns. Each number in a matrix is an element or entry. The plural of matrix is matrices. Example: $\begin{pmatrix} 2 & 3 \\ 0 & -1 \end{pmatrix}$ is a matrix with 4 elements.

Monomial — a number, a variable, or the product of numbers and variables with or without exponents. Examples: $3x$, 7, $2xy$, $2y^2$, x^3

Polynomial — a monomial or the sum of monomials. Each monomial is called a term of the polynomial. Example: $3x^2 - 7 + 2x + 5y - 2y^2 + x^3$

Pythagorean Theorem — a statement of the relationship between the lengths of the sides in a right triangle. If a and b are legs, and c is the hypotenuse, $a^2 + b^2 = c^2$.

Quadratic Equation — a polynomial in the form of $ax^2 + bx + c$, where $a \neq 0$.

Radical Expression — an expression that contains a radical, such as a square root or cubed root.

Rational Number — a number that can be written as the ratio of two whole numbers. Example: 7 is rational; it can be written as $\frac{7}{1}$; 0.25 is rational; it can be written as $\frac{1}{4}$.

Slope — the ratio of the *rise* (vertical change) to the *run* (horizontal change) for a non-vertical line.

Square Root — a number that when multiplied by itself gives you another number. The symbol for square root is $\sqrt{x}$. Example: $\sqrt{49} = 7$ reads "The square root of 49 is 7."

Straight Angle — an angle measuring exactly 180°.

Supplementary Angles — two angles whose measures add up to 180°.

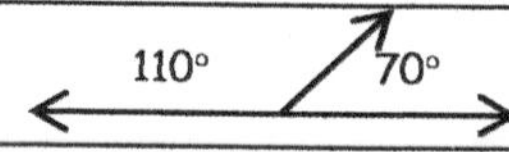

Surface Area — the sum of the areas of all of the faces of a solid figure.

Term — the components of an expression, usually being added to or subtracted from each other. Example: The expression $2x + 5$ has two terms: $2x$ and 5. The expression $3n^2$ has only one term.

Trinomial — a polynomial having exactly 3 terms. Examples: $3x - 7 + 2x$, $5y + 2y^2 - x^3$

Help Pages

Vocabulary (continued)

Geometry — Circles

Circumference — the distance around the outside of a circle.

Diameter — the widest distance across a circle. The diameter always passes through the center.

Radius — the distance from any point on the circle to the center. The radius is half of the diameter.

Geometry — Triangles

Equilateral — a triangle with all 3 sides having the same length.

Isosceles — a triangle with 2 sides having the same length.

Scalene — a triangle with none of the sides having the same length.

Geometry — Polygons

Number of Sides	Name	Number of Sides	Name
3	Triangle	7	Heptagon
4	Quadrilateral	8	Octagon
5	Pentagon	9	Nonagon
6	Hexagon	10	Decagon

Measurement — Relationships

Volume	Distance
3 teaspoons in a tablespoon	36 inches in a yard
2 cups in a pint	1760 yards in a mile
2 pints in a quart	5280 feet in a mile
4 quarts in a gallon	100 centimeters in a meter
Weight	1000 millimeters in a meter
16 ounces in a pound	**Temperature**
2000 pounds in a ton	0° Celsius - Freezing Point
Time	100°Celsius - Boiling Point
10 years in a decade	32°Fahrenheit - Freezing Point
100 years in a century	212°Fahrenheit - Boiling Point

Help Pages

Solved Examples

Absolute Value

The **absolute value** of a number is its distance from zero on a number line. It is always positive.

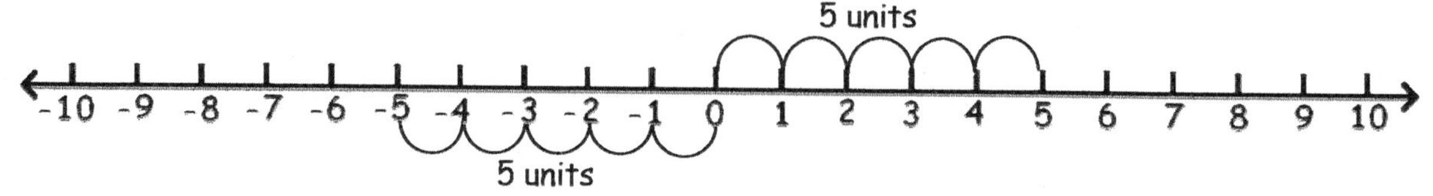

The absolute value of both -5 and +5 is 5, because both are 5 units away from zero. The symbol for the absolute value of -5 is |-5|. Examples: |-3| = 3; |8| = 8.

Equations

More complicated equations involve variables which replace a number. To solve an equation like this, you must figure out which number the variable stands for. There is a process for solving for the variable. No matter how complicated the equation, the goal is to work with the equation until all the numbers are on one side and the variable is alone on the other side. To check your answer, put the value of x back into the original equation.

These multi-step equations also have a variable on only one side. To get the variable alone, though, requires several steps.

Example: Solve for x. $3(2x+3)=21$

$$\cancel{3}\left(\frac{2x+3}{\cancel{3}}\right)=\frac{21}{3}$$

$$2x+3=7$$

$$-3=-3$$

$$2x=4$$

$$\frac{\cancel{2}x}{\cancel{2}}=\frac{4}{2}$$

$$x=2$$

Check: $3(2(2)+3)=21$

$3(4+3)=21$

$3(7)=21$

$21 = 21$ ✓ correct!

1. Look at the side of the equation that has the variable on it. First, $(2x+3)$ is multiplied by 3; then there is a number (3) added to $2x$, and there is a number (2) multiplied by x. All of these must be removed. To remove the 3 outside the parentheses, divide both sides by 3.
2. To remove the 3 inside the parentheses, add its opposite (-3) to both sides.
3. Remove the 2 by dividing both sides by 2. $2x$ divided by 2 is x. 4 divided by 2 is two.
4. Once the variable is alone on one side of the equation, the equation is solved. The last step tells the value of x. $x = 2$.

When solving an **equation with a variable on both sides**, the goals are the same: to get the numbers on one side of the equation and to get the variable alone on the other side.

Example: Solve for x. $2x + 4 = 6x - 4$

$$2x+4=6x-4$$

$$-2x\quad=-2x$$

$$4=4x-4$$

$$+4=\quad+4$$

$$8=4x$$

$$\frac{8}{4}=\frac{\cancel{4}x}{\cancel{4}}$$

$$2=x$$

1. Since there are variables on both sides, the first step is to remove the "variable term" from one of the sides. To remove $2x$ from the left side, add $-2x$ to both sides.
2. Next, remove the number added to the variable side by adding its opposite. To remove -4, add +4 to both sides.
3. The variable still has a number multiplied by it (4), which can be removed by dividing both sides by 4.

Solved Examples

Equations (continued)

A **quadratic equation** is a polynomial in the form $ax^2 + bx + c = 0$, where $a \neq 0$. Some quadratic equations can be solved simply by factoring and then setting them equal to zero.

Example: Solve the equation for x. $x^2 + 5x + 6 = 0$

$x^2 + 5x + 6 = (x+2)(x+3)$ ← 1. First, factor the quadratic. (See p. 304)

$(x+2)(x+3) = 0$ ← 2. Next, substitute the factors into the equation.

$(x+2) = 0,\ x = -2$
$(x+3) = 0,\ x = -3$ ← 3. To solve for x, realize that one or both of the factors must equal zero.

$x = \{-2, -3\}$ ← 4. The values of x that make the factors equal zero are the solutions.

Sometimes a quadratic equation can't be factored. In that case, the method described above is not useful; in this case, the **quadratic formula** must be used.

For quadratic equations in the standard form $ax^2 + bx + c = 0$,

the quadratic formula is $x = \dfrac{-b \pm \sqrt{b^2 - 4ac}}{2a}$, where $a \neq 0$ and $b^2 - 4ac \geq 0$.

Example 1: Solve the equation $3x^2 + 5x = 8$ for x.

$3x^2 + 5x = 8$

$3x^2 + 5x - 8 = 0$ ← 1. Rewrite the equation in standard form. Remember, in standard form, it must equal 0.

$ax^2 + bx + c = 0$

$a = 3,\ b = 5,\ c = -8$ ← 2. Identify the values of a, b, c in the equation.

$x = \dfrac{-b \pm \sqrt{b^2 - 4ac}}{2a}$

$x = \dfrac{-5 \pm \sqrt{5^2 - 4(3)(-8)}}{2(3)}$ ← 3. Put the values of a, b, c into the quadratic formula.

4. Simplify each part of the fraction.

$x = \dfrac{-5 \pm \sqrt{121}}{6}$

$x = \dfrac{-5 + 11}{6}$ and $\dfrac{-5 - 11}{6}$

$x = \dfrac{6}{6} = 1$ and $\dfrac{-16}{6} = -\dfrac{8}{3}$ ← 5. Because of the $\pm$ in the formula, there will be two solutions.

Example 2: Solve for n. $n^2 - 7n + 10 = 0$ $a = 1,\ b = -7,\ c = 10$

$x = \dfrac{-(-7) \pm \sqrt{(-7)^2 - 4(1)(10)}}{2(1)}$

$x = \dfrac{+7 \pm \sqrt{49 - 40}}{2}$

$x = \dfrac{7 \pm \sqrt{9}}{2}$

$x = \dfrac{7 \pm 3}{2}$ → $x = \dfrac{7+3}{2} = \dfrac{10}{2} = 5$ and $x = \dfrac{7-3}{2} = \dfrac{4}{2} = 2$

The solutions are 5 and 2.

Solved Examples

Equations (continued)

When solving any equation, there can only be one unknown variable. Sometimes there are multiple variables that are not known. When this is the case, the only way to solve is to use a **system of equations**. A system of equations is a group of equations, all having the same unknown variables. There must be the same number of equations as there are unknown variables. For example, if there are 2 unknown variables, the system must include 2 equations; if there are 3 unknown variables, the system must include 3 equations.

We will use 2 different methods for solving systems of equations: Substitution and Elimination.

To **solve a system of equations by substitution**, there are 3 simple steps:

1. First choose one of the equations, and use it to solve for one of the variables. (Usually this variable will be equal to an expression in terms of the other variable).
2. Use the expression from step 1 and replace that first variable in the second equation. Solve the second equation for the other variable.
3. Substitute the value from step 2 into the first equation and solve. You now know the values of both variables.

Example: Use substitution to solve the system for x and y.

$y-2=3x$ Equation 1
$x+2y=11$ Equation 2

1. Using Equation 1, solve for y. (You now have an expression that is equal to y.)
2. Using Equation 2, substitute this expression in place of y. Now the equation only has one variable left, x. Solve for x.
3. Now that you know the value of x, go back to Equation 1 and substitute the value of x into the equation. Solve for y.

$y-2=3x$ Eqn.1
$y=3x+2$

$x+2y=11$ Eqn. 2
$x+2(3x+2)=11$
$x+6x+4=11$
$7x+4=11$
$7x=7$
$x=1$

$y-2=3(1)$ Eqn. 1
$y=3+2$
$y=3+2=5$

The solution to this system is x= 1 and y= 5.

To **solve a system of equations by elimination**, there are also 3 simple steps:

1. Add or subtract the equations to eliminate one of the variables.
2. Solve the resulting equation for the remaining variable.
3. Substitute the value back into either equation. Solve for the value of the eliminated variable.

Example: Use elimination to solve the system for x and y.

$2x+3y=11$ Equation 1
$4x-3y=13$ Equation 2

1. Add the equations to eliminate one variable, y.
2. Solve for the other variable, x.
3. Substitute the value of x into one of the original equations and solve for y.

$2x+3y=11$
$+\ 4x-3y=13$
$6x \quad = 24$
$x \quad = 4$

$2x+3y=11$
$2(4)+3y=11$
$8+3y=11$
$3y=3$
$y=1$

The solution to this system is x= 4 and y= 1.

Help Pages

Solved Examples

Exponents

You are very familiar with exponents and their meaning. Until now, though, exponents have been positive numbers. What if the **exponent is zero or a negative** number? The rules for dealing with **negative** or **zero exponents** are as follows

a to the zero power is 1. $a^0 = 1, a \neq 0,$ so $7^0 = 1$

a^{-n} is the reciprocal of a^n. $a^{-n} = \frac{1}{a^n}, a \neq 0,$ so $4^{-1} = \frac{1}{4}$

a^n is the reciprocal of a^{-n}. $\frac{1}{a^{-n}} = a^n, a \neq 0,$ so $\frac{1}{9^{-1}} = 9^1 = 9$

Examples: $(-10)^0 = 1$ $\left(\frac{1}{4}\right)^{-2} = \frac{1}{\left(\frac{1}{4}\right)^2} = \frac{1}{\left(\frac{1}{16}\right)} = 16$ $\frac{1}{8^{-2}} = 8^2 = 64$

When **multiplying exponential terms** that have the same base, keep the base and add the exponents.

Examples: $a^2 \cdot a^3 = a^{2+3} = a^5$ $x^4 \cdot x^5 = x^{4+5} = x^9$

When **dividing exponential terms** with the same base, keep the base and subtract the exponents.

Example 1: $\frac{a^6}{a^2} = \frac{a \cdot a \cdot a \cdot a \cdot \not{a} \cdot \not{a}}{\not{a} \cdot \not{a}} = a \cdot a \cdot a \cdot a = a^4 = a^{6-2}$

Example 2 $\frac{b^{10}}{b^3} = b^{10-3} = b^7$

Sometimes the entire quotient is raised to a power. In that case, apply the exponent to both the numerator and denominator, simplify each of them, and then divide, if possible.

Example 1: $\left(\frac{2x^2}{5y}\right)^3 = \frac{(2x^2)^3}{(5y)^3} = \frac{2^3 \cdot (x^2)^3}{5^3 \cdot y^3} = \frac{8x^6}{125y^3}$

Example 2: $\left(\frac{3x^2}{9y^2}\right)^2 = \frac{(3x^2)^2}{(9y^2)^2} = \frac{3^2 \cdot (x^2)^2}{9^2 \cdot (y^2)^2} = \frac{\not{9}x^4}{\underset{9}{\not{81}}y^4} = \frac{x^4}{9y^4}$

When **raising an exponential term to a power**, keep the base and multiply the exponents.

Examples: $(a^2)^3 = a^{2\times3} = a^6$ $(x^4)^4 = x^{4\times4} = x^{16}$

Expressions

An **expression** is a number, a variable, or any combination of these, along with operation signs $(+, -, \times, \div)$ and grouping symbols. An expression never includes an equal sign.

Five examples of expressions are 5, x, $(x + 5)$, $(3x + 5)$, and $(3x^2 + 5)$.

To **evaluate an expression** means to calculate its value using specific variable values.

Example: Evaluate $2x + 3y + 5$ when $x = 2$ and $y = 3$.

$2(2) + 3(3) + 5 = ?$

$4 + 9 + 5 = ?$

$13 + 5 = 18$

The expression has a value of 18.

1. To evaluate, put the values of x and y into the expression.
2. Use the rules for integers to calculate the value of the expression.

Help Pages

Solved Examples

Expressions (continued)

Some expressions can be simplified. There are a few processes for **simplifying an expression.** Deciding which process or processes to use depends on the expression itself. With practice, you will be able to recognize which of the following processes to use.

The **distributive property** is used when one term is multiplied by (or divided into) an expression that includes either addition or subtraction. $a(b+c)=ab+ac$ or $\frac{b+c}{a}=\frac{b}{a}+\frac{c}{a}$

Example: Simplify. $3(2x+5)$

$$3(2x+5)=$$
$$3(2x)+3(5)=$$
$$6x+15$$

1. Since the 3 is multiplied by the expression, $2x+5$, the 3 must be multiplied by both terms in the expression.
2. Multiply 3 by $2x$ and then multiply 3 by +5.
3. The result includes both of these: $6x+15$.

Notice that simplifying an expression does not result in a single number answer, only a simpler expression.

Example: Simplify. $\frac{14x-6y+8}{2}$

$$\frac{14x-6y+8}{2}=$$
$$\frac{14x}{2}-\frac{6y}{2}+\frac{8}{2}=$$
$$7x-3y+4$$

Expressions which contain like terms can also be simplified. **Like terms** are those that contain the same variable to the same power. $2x$ and $-4x$ are like terms; $3n^2$ and $8n^2$ are like terms; $5y$ and y are like terms; 3 and 7 are like terms.

An expression sometimes begins with like terms. This process for **simplifying expressions** is called **combining like terms**. When combining like terms, first identify the like terms. Then, simply add the like terms to each other and write the results together to form a new expression.

Example: Simplify. $2x+5y-9+5x-3y-2$.

The like terms are $2x$ and $+5x$, $+5y$ and $-3y$, and -9 and -2.

$2x + +5x =$ $\mathbf{+7x}$, $+5y + -3y =$ $\mathbf{+2y}$, and $-9 + -2 =$ $\mathbf{-11}$.

The result is $\mathbf{7x + 2y - 11}$.

The next examples are a bit more complex. It is necessary to use the distributive property first, and then to combine like terms.

Example: Simplify. $2(3x+2y+2)+3(2x+3y+2)$

$$\begin{array}{r} 6x+4y+4 \\ +6x+9y+6 \\ \hline 12x+13y+10 \end{array}$$

Example: Simplify. $4(3x-5y-4)-2(3x-3y+2)$

$$\begin{array}{r} +12x-20y-16 \\ -6x+6y-4 \\ \hline 6x-14y-20 \end{array}$$

1. First, apply the distributive property to each expression. Write the results on top of each other, lining up the like terms with each other. Pay attention to the signs of the terms.
2. Then, add each group of like terms. Remember to follow the rules for integers.

Help Pages

Solved Examples

Expressions (continued)

Other expressions that can be simplified are written as fractions. **Simplifying** these expressions (**algebraic fractions**) is similar to simplifying numerical fractions. It involves cancelling out factors that are common to both the numerator and the denominator.

Simplify. $\dfrac{12x^2yz^4}{16xy^3z^2}$

$$\frac{\overset{3}{\cancel{12}}\,\overset{x}{\cancel{x^2}}\,\cancel{y}\,\overset{z^2}{\cancel{z^4}}}{\underset{4}{\cancel{16}}\,\cancel{x}\,\underset{y^2}{\cancel{y^3}}\,\cancel{z^2}}$$

$$\frac{\cancel{2}\cdot\cancel{2}\cdot 3\cdot\cancel{x}\cdot x\cdot\cancel{y}\cdot\cancel{z}\cdot\cancel{z}\cdot z\cdot z}{\cancel{2}\cdot\cancel{2}\cdot 2\cdot 2\cdot\cancel{x}\cdot\cancel{y}\cdot y\cdot y\cdot\cancel{z}\cdot\cancel{z}}$$

$$\frac{3xz^2}{4y^2}$$

1. Begin by looking at the numerals in both the numerator and denominator (12 and 16). What is the largest number that goes into both evenly? Cancel this factor (4) out of both.
2. Look at the x portion of both numerator and denominator. What is the largest number of x's that can go into both of them? Cancel this factor (x) out of both.
3. Do the same process with y and then z. Cancel out the largest number of each (y and z^2). Write the numbers that remain in the numerator or denominator for your answer.

Often a relationship is described using verbal (English) phrases. In order to work with the relationship, you must first **translate it into an algebraic expression or equation**. In most cases, word clues will be helpful. Some examples of verbal phrases and their corresponding algebraic expressions or equations are written below.

Verbal Phrase	Algebraic Expression/Equation
Ten more than a number	$x + 10$
The sum of a number and five	$x + 5$
A number increased by seven	$x + 7$
Six less than a number	$x - 6$
A number decreased by nine	$x - 9$
The difference between a number and four	$x - 4$
The difference between four and a number	$4 - x$
Five times a number	$5x$
Eight times a number, increased by one	$8x + 1$
The quotient of a number and 10	$\frac{x}{10}$
The quotient of a number and two, decreased by five	$\frac{x}{2} - 5$
The product of a number and six is twelve.	$6x = 12$

In most problems, the word "is" tells you to put in an equal sign. When working with fractions and percents, the word "of" generally means multiply. Look at the example below.

One half <u>of</u> a number <u>is</u> fifteen.

You can think of it as "one half <u>times</u> a number <u>equals</u> fifteen."

When written as an algebraic equation, it is $\frac{1}{2}x = 15$.

Help Pages

Solved Examples

Expressions (continued)

At times you need to find the **Greatest Common Factor (GCF) of an algebraic expression.**

Example: Find the GCF of $12x^2yz^3$ and $18xy^3z^2$.

1. First, find the GCF of the numbers (12 and 18). The largest number that is a factor of both is **6**.
2. Now look at the x's. Of the x-terms, which contains fewer x's. Comparing x^2 and x, $\boldsymbol{x}$ contains fewer.
3. Now look at the y's and then the z's. Again, of the y-terms, $\boldsymbol{y}$ contains fewer. Of the z-terms, $\boldsymbol{z^2}$ contains fewer.
4. The GCF contains all of these: $\boldsymbol{6xyz^2}$.

$12x^2yz^3$ and $18xy^3z^2$

The GCF of 12 and 18 is **6**.

Of x^2 and x, the smaller is $\boldsymbol{x}$.

Of y and y^3, the smaller is $\boldsymbol{y}$.

Of z^3 and z^2, the smaller is $\boldsymbol{z^2}$.

The GCF is: $\boldsymbol{6xyz^2}$.

At other times you need to know the **Least Common Multiple (LCM) of an algebraic expression.**

Example: Find the LCM of $10a^3b^2c^2$ and $15ab^4c$.

1. First, find the LCM of the numbers (10 and 15). The lowest number that both go into evenly is **30**.
2. Now look at the a-terms. Which has the larger number of a's? Comparing a^3 and a, $\boldsymbol{a^3}$ has more.
3. Now look at the b's and then the c's. Again, of the b-terms, $\boldsymbol{b^4}$ contains more. Of the c-terms, $\boldsymbol{c^2}$ contains more.
4. The LCM contains all of these: $\boldsymbol{30a^3b^4c^2}$.

$10a^3b^2c^2$ and $15ab^4c$

The LCM of 10 and 15 is **30**.

Of a^3 and a, the larger is $\boldsymbol{a^3}$.

Of b^2 and b^4, the larger is $\boldsymbol{b^4}$.

Of c^2 and c, the larger is $\boldsymbol{c^2}$.

The LCM is: $\boldsymbol{30a^3b^4c^2}$.

When **adding (subtracting) rational expressions**, the process is similar to that used to add (subtract) ordinary fractions. Just as with fractions, rational expressions must have a common denominator before they can be added (subtracted).

Example 1: Add. $\frac{5}{3x}+\frac{7}{3x}$

$$\frac{5}{3x}+\frac{7}{3x}=\frac{\cancel{12}}{\cancel{3}x}=\frac{4}{x}$$

1. Since the expressions already have a common denominator, they can be added.
2. Add the numerators; keep the denominator.
3. Simplify.

Example 2: Subtract. $\frac{x}{2}-\frac{(x-1)}{(x-2)}$

$2(x-2)$ is the common denominator.

$$\cancel{2}(x-2)\frac{x}{\cancel{2}}-2\cancel{(x-2)}\frac{(x-1)}{\cancel{(x-2)}}=$$

$$(x-2)x-2(x-1)=$$

$$x^2-2x-2x+2=$$

$$x^2-4x+2=$$

$$(x-2)(x-2)$$

1. The expressions need a common denominator, before they can be added. Multiply the denominators to get a common denominator.
2. Multiply each fraction by the common denominator. Cancel where you can.
3. Simplify and combine like terms.
4. Simplify by factoring.

Solved Examples

Expressions (continued)

When **multiplying (dividing) rational expressions**, the process is similar to that used to multiply (divide) ordinary fractions. A common denominator is not necessary to multiply (divide) rational expressions.

Example 1: Multiply. $\frac{2x^2}{3x} \cdot \frac{6x^2}{12x^3}$

$$\frac{\cancel{12}\,\cancel{x^4}}{\cancel{36}\,\cancel{x^4}} =$$

$$\frac{1}{3}$$

1. Just as with ordinary fractions, multiply the numerators; multiply the denominators.
2. Cancel where possible.
3. Simplify.

Example 2: Divide. $\frac{7x^2 - 7x}{x^2 + 2x - 3} \div \frac{x+1}{x^2 - 7x - 8}$

$$\frac{7x^2 - 7x}{x^2 + 2x - 3} \times \frac{x^2 - 7x - 8}{x+1} =$$

$$\frac{(7x^2 - 7x)(x^2 - 7x - 8)}{(x^2 + 2x - 3)(x+1)} =$$

$$\frac{7x\cancel{(x-1)}(x-8)\cancel{(x+1)}}{(x+3)\cancel{(x-1)}\cancel{(x+1)}} =$$

$$\frac{7x(x-8)}{(x+3)}$$

1. Just as with ordinary fractions, to divide, take the reciprocal of the 2nd fraction and multiply.
2. Multiply the numerators; multiply the denominators.
3. Simplify the expressions where possible by factoring.
4. Cancel where possible.
5. Simplify.

Functions

A function is a rule that pairs each number in a given set (the domain) with just one number in another set (the range). A function performs one or more operations on an input-number which results in an output-number. The set of all input-numbers is called the **domain** of the function. The set of all output-numbers is called the **range** of the function. Often, a function table is used to help organize your thinking.

Example: For the function, $y = 3x$, find the missing numbers in the function table.

x	y
2	?
-1	?
?	15

The function is $y = 3x$. This function multiplies every x-value by 3.

When we input $x = 2$, we get $y = 3(2)$ or $y = 6$. →

When we use $x = -1$, we get $y = 3(-1)$ or $y = -3$. →

When we use $y = 15$, we get $15 = 3x$, so $\frac{15}{3} = x$ or $5 = x$. →

x	y
2	6
-1	-3
5	15

The set of all inputs is the domain. For this function table, the domain is {2, -1, 5}.

The set of all outputs is the range. For this function table, the range is {6, -3, 15}.

Help Pages

Solved Examples

Geometry

Finding the **area of a parallelogram** is similar to finding the area of any other quadrilateral. The area of the figure is equal to the length of its base multiplied by the height of the figure.

Area of parallelogram = base × height or $A = b \times h$

Example: Find the area of the parallelogram below.

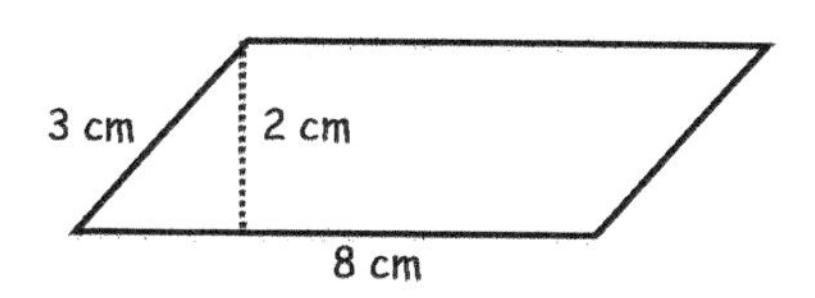

1. Find the length of the base. (8 cm)
2. Find the height. (It is 2cm. The height is always straight up and down – never slanted.)
3. Multiply to find the area. (16 cm^2)

So, $A = 8\text{ cm} \times 2\text{ cm} = \mathbf{16\text{ cm}^2}$.

Finding the **area of a trapezoid** is a little different than the other quadrilaterals that we have seen. Trapezoids have 2 bases of unequal length. To find the area, first find the average of the lengths of the 2 bases. Then, multiply that average by the height.

Area of trapezoid = $\frac{\text{base}_1 + \text{base}_2}{2} \times \text{height}$ or $A = (\frac{b_1 + b_2}{2})h$

Example: Find the area of the trapezoid below.

1. Add the lengths of the two bases. (22 cm)
2. Divide the sum by 2. (11 cm)
3. Multiply that result by the height to find the area. (110 cm^2)

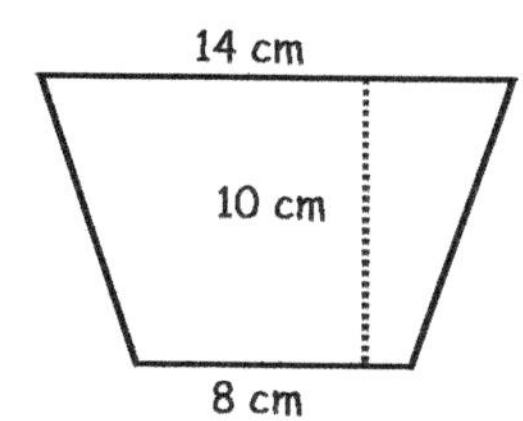

$$\frac{14\text{cm} + 8\text{cm}}{2} = \frac{22\text{cm}}{2} = 11\text{cm}$$

$11\text{ cm} \times 10\text{ cm} = \mathbf{110\text{ cm}^2}$ = Area

To find the **area of a triangle**, first recognize that any triangle is exactly half of a parallelogram.

The whole figure is a parallelogram.

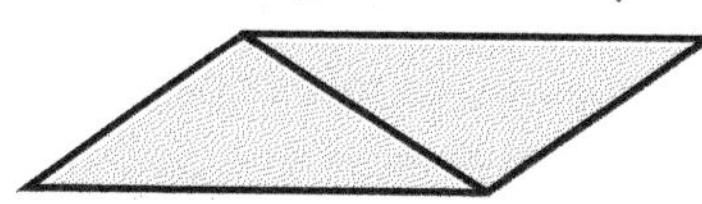

Half of the whole figure is a triangle.

So, the triangle's area is equal to half of the product of the base and the height.

Area of triangle = $\frac{1}{2}$(base × height) or $A = \frac{1}{2}bh$

Examples: Find the area of the triangles below.

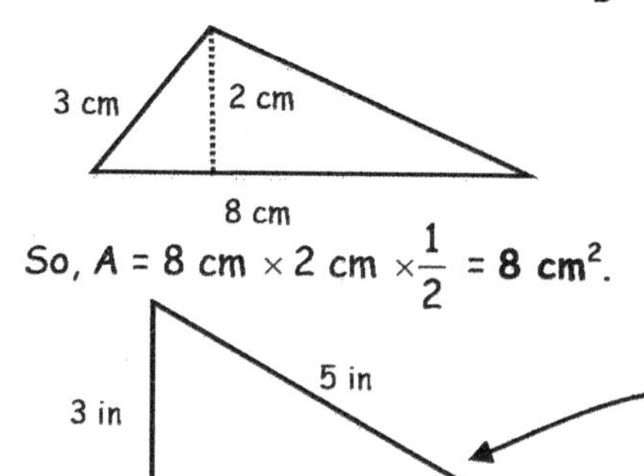

So, $A = 8\text{ cm} \times 2\text{ cm} \times \frac{1}{2} = \mathbf{8\text{ cm}^2}$.

1. Find the length of the base. (8 cm)
2. Find the height. (It is 2cm. The height is always straight up and down – never slanted.)
3. Multiply them together and divide by 2 to find the area. (8 cm^2)

3 in

5 in

4 in

The base of this triangle is 4 inches long. Its height is 3 inches. (Remember the height is always straight up and down!)

So, $A = 4\text{ in} \times 3\text{ in} \times \frac{1}{2} = \mathbf{6\text{ in}^2}$.

Solved Examples

Geometry (continued)

Remember there are three types of triangles: acute, obtuse, and right. **A right triangle** has one 90° angle. (The other 2 angles will be less than 90°; the sum of the angles in any triangle is 180°.)

Every triangle has 3 sides; in a right triangle, the sides have names. There are two legs and a hypotenuse. The legs are the sides that come together to form the right angle; they are said to be adjacent to the right angle. The hypotenuse is the side that is opposite the right angle.

In this right triangle, sides **a** and **b** are the legs.

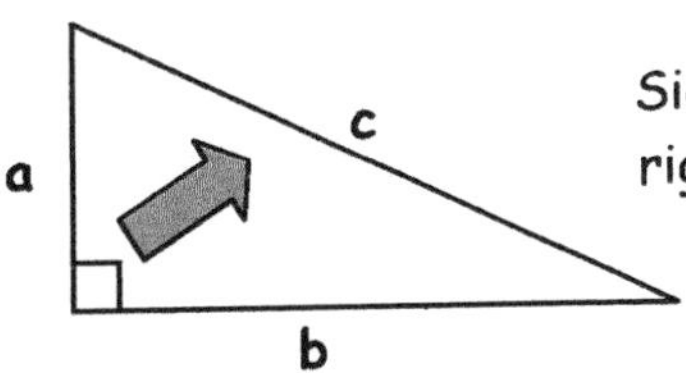

Side **c** is across from (opposite) the right angle. Side **c** is the hypotenuse.

Right triangles have many special relationships. Here, the focus will be on one relationship in particular – the relationship defined by the **Pythagorean Theorem**.

The Pythagorean Theorem states "In any right triangle, the sum of the squares of the legs is equal to the square of the hypotenuse." To put it another way, $\mathbf{a^2 + b^2 = c^2}$.

Example 1: Find the measure of the third side in this triangle.

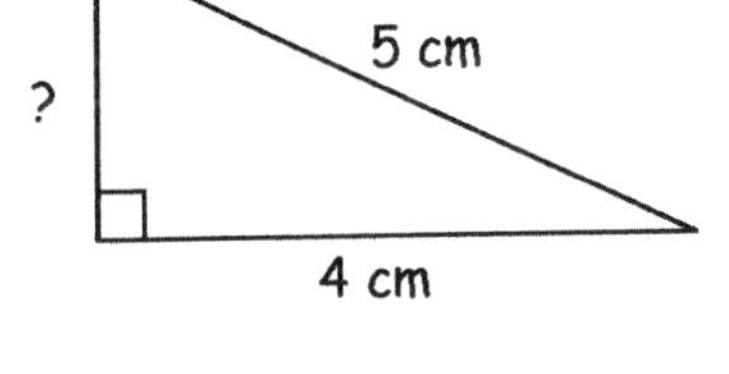

$a^2 + b^2 = c^2$

(Note that the missing side is a leg.)

$a^2 + 4^2 = 5^2$

$a^2 + 16 = 25$

$a^2 = 9$

$a = \sqrt{9} = 3$ (Remember: To undo a square, find the square root.)

Example 2: Find the measure of the third side in this triangle.

5 cm

?

6 cm

$a^2 + b^2 = c^2$

(Note that the missing side is the hypotenuse.)

$5^2 + 6^2 = c^2$

$25 + 36 = c^2$

$61 = c^2$

$\sqrt{61} = c$

Solved Examples

Geometry (continued)

The **circumference of a circle** is the distance around the outside of the circle. Before you can find the circumference of a circle you must know either its radius or its diameter. Also, you must know the value of the constant, *pi* (π). $\pi = 3.14$ (rounded to the nearest hundredth).

Once you have this information, the circumference can be found by multiplying the diameter by *pi.*

Circumference = $\pi \times$ diameter or $C = \pi d$

Examples: Find the circumference of the circles below.

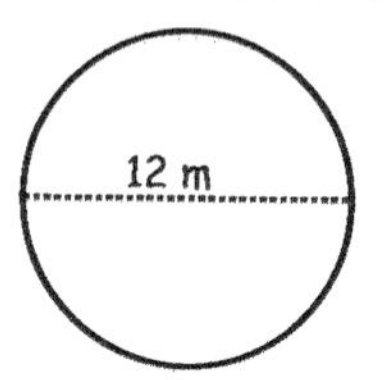

1. Find the length of the diameter. (12 m)
2. Multiply the diameter by π. (12m $\times$ 3.14)
3. The product is the circumference. (37.68 m)

So, C = 12 m $\times$ 3.14 = **37.68 m.**

Sometimes the radius of a circle is given instead of the diameter. Remember, the radius of any circle is exactly half of the diameter. If a circle has a radius of 3 feet, its diameter is 6 feet.

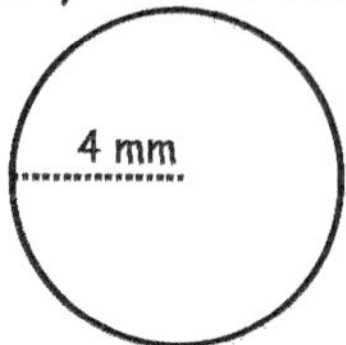

Since the radius is 4 mm, the diameter must be 8 mm.
Multiply the diameter by π. (8 mm $\times$ 3.14)
The product is the circumference. (25.12 mm)

So, C = 8 mm $\times$ 3.14 = **25.12 mm.**

When finding the **area of a circle**, the length of the radius is squared (multiplied by itself), and then that answer is multiplied by the constant, *pi* (π). $\pi = 3.14$ (rounded to the nearest hundredth).

Area = $\pi \times$ radius $\times$ radius or $A = \pi r^2$

Examples: Find the area of the circles below.

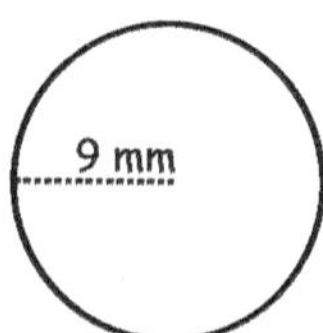

1. Find the length of the radius. (9 mm)
2. Multiply the radius by itself. (9 mm x 9 mm)
3. Multiply the product by *pi.* (81 mm^2 x 3.14)
4. The result is the area. (254.34 mm^2)

So, A = 9 mm x 9 mm x 3.14 = **254.34 mm^2.**

Sometimes the diameter of a circle is given instead of the radius. Remember, the diameter of any circle is exactly twice the radius. If a circle has a diameter of 6 feet, its radius is 3 feet.

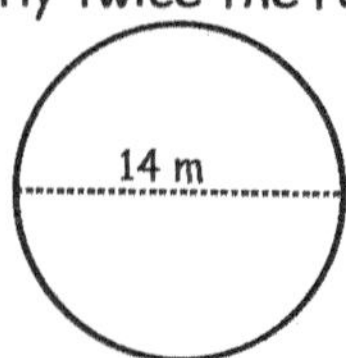

Since the diameter is 14 m, the radius must be 7 m.
Square the radius. (7 m x 7 m)
Multiply that result by π. (49 m^2 $\times$ 3.14)
The product is the area. (153.86 m^2)

So, A = (7 m)^2 $\times$ 3.14 = **153.86 m^2.**

Help Pages

Solved Examples

Geometry (continued)

To find the **surface area** of a solid figure, it is necessary to first count the total number of faces. Then, find the area of each of the faces; finally, add the areas of each face. That sum is the surface area of the figure.

Here, the focus will be on finding the **surface area of a rectangular prism**. A rectangular prism has 6 faces. Actually, the opposite faces are identical, so this figure has 3 pairs of faces. Also, a prism has only 3 dimensions: Length, Width, and Height.

This prism has identical left and right sides (A & B), identical top and bottom (C & D), and identical front and back (unlabeled).

1. Find the area of the front: L x W. (10 m x 5 m = 50 m^2) Since the back is identical, its area is the same.
2. Find the area of the top (C): L x H. (10 m x 2 m = 20 m^2) Since the bottom (D) is identical, its area is the same.
3. Find the area of side A: W x H. (2 m x 5 m = 10 m^2) Since side B is identical, its area is the same.
4. Add up the areas of all 6 faces. (10 m^2 + 10 m^2 + 20 m^2 + 20 m^2 + 50 m^2+ 50 m^2 = **160 m^2**)

Surface Area of a Rectangular Prism = 2(length x width) + 2(length x height) + 2(width x height)

or SA = 2LW + 2LH + 2WH

To find the **volume** of a solid figure, it is necessary to determine the area of one face and multiply that by the height of the figure. Volume of a solid is measured in cubic units (cm^3, in^3, ft^3, etc.).

Here the focus will be on finding the **volume of a cylinder**. As shown below, a cylinder has two identical circular faces.

Example: Find the volume of the cylinder below.

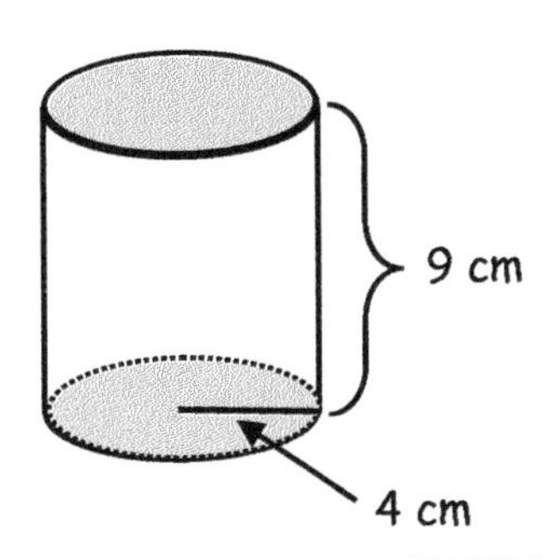

1. To find the area of one of the circular faces, multiply the constant, π (3.14), by the square of the radius (4 cm). Area = 3.14 × $(4\text{ cm})^2$ = 50.24 cm^2
2. The height of this cylinder is 9 cm. Multiply the height by the area calculated in Step 1.
   Volume = 50.24 cm^2 × 9 cm = 452.16 cm^3

Help Pages

Solved Examples

Graphing

A **coordinate plane** is formed by the intersection of a horizontal number line, called the ***x*-axis**, and a vertical number line, called the ***y*-axis**. The axes meet at the point (0, 0), called the **origin**, and divide the coordinate plane into four **quadrants**.

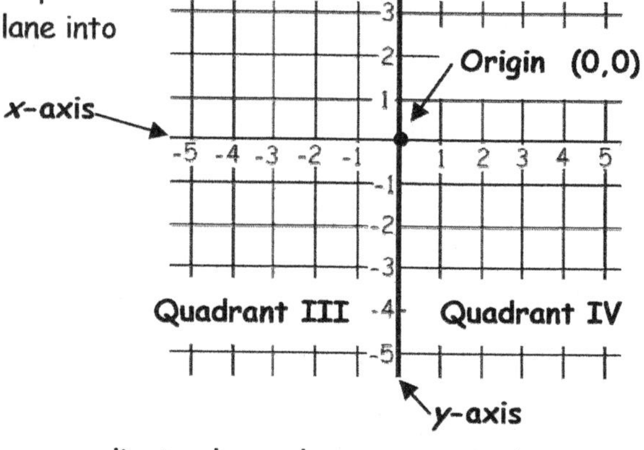

Points are represented by **ordered pairs** of numbers, (x, y). The first number in an ordered pair is the x-coordinate; the second number is the y-coordinate. In the point **(-4, 1)**, -4 is the x-coordinate and 1 is the y-coordinate.

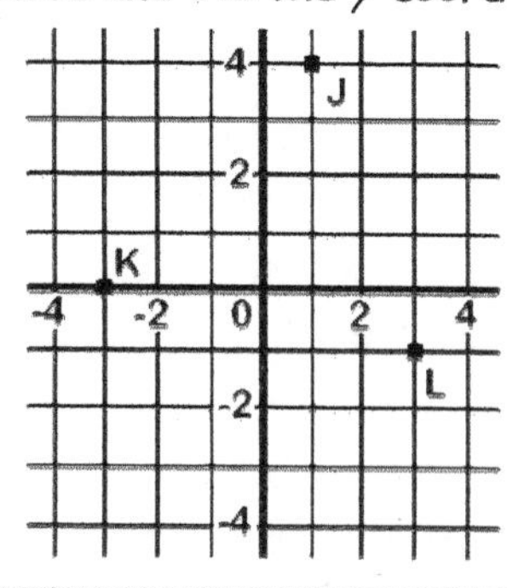

When graphing on a coordinate plane, always move on the x-axis first (right or left), and then move on the y-axis (up or down).

- The coordinates of point J are (1, 4).
- The coordinates of point K are (-3, 0).
- The coordinates of point L are (3, -1).

On a coordinate plane, any 2 points can be connected to form a line. The line, however, is made up of many points – in fact, every place on the line is another point. One of the properties of a line is its slope (or steepness). The **slope** of a non-vertical line is the ratio of its vertical change (rise) to its horizontal change (run) between any two points on the line. The slope of a line is represented by the letter m. Another property of a line is the ***y*-intercept**. This is the point where the line intersects the y-axis. A line has only one y-intercept, which is represented by the letter b.

$$\text{Slope of a line} = \frac{\text{change in } y}{\text{change in } x} = \frac{\text{rise}}{\text{run}}$$

The rise-over-run method can be used to find the slope if you're looking at the graph.

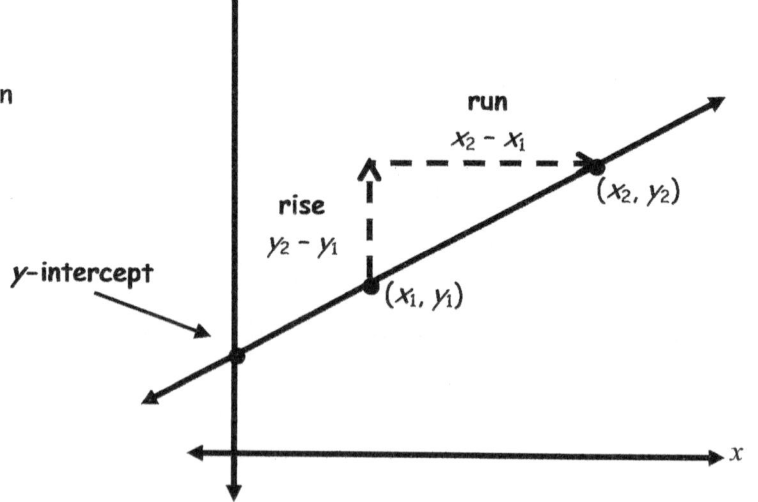

Solved Examples

Graphing (continued)

Another way to find the slope of a line is to use the formula. The formula for slope is $m = \frac{y_2 - y_1}{x_2 - x_1}$, where the two points are (x_1, y_1) and (x_2, y_2).

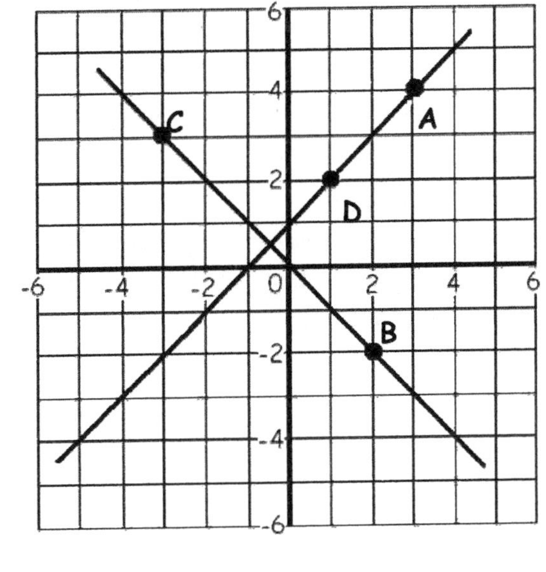

Example: What is the slope of $\overline{AD}$?

Point A with coordinates (3, 4) and Point D with coordinates (1, 2) are both on this line.

For Point A, x_2 is 3 and y_2 is 4.

For Point D, x_1 is 1 and y_1 is 2.

slope = $m = \frac{y_2 - y_1}{x_2 - x_1} = \frac{4-2}{3-1} = \frac{2}{2} = 1$

The slope of $\overline{AD}$ is 1.

Use the formula to find the slope of $\overline{CB}$.

Point C is (-3, 3) and Point B is (2, -2).

slope = $m = \frac{y_2 - y_1}{x_2 - x_1} = \frac{-2-3}{2-(-3)} = \frac{-5}{5} = -1$

The slope of $\overline{CB}$ is -1.

Every line has an equation which describes it, called a linear equation. We will focus on one particular form of linear equation – **slope-intercept form**. To write the slope-intercept equation of a line, you must know the slope and the y-intercept.

A linear equation in slope-intercept form is always in the form $y = mx + b$, where m is the slope, b is the y-intercept, and (x, y) is any point on the line.

Example: A line has the equation $y = 2x + 5$. What is the slope? What is the y-intercept?

$$y = 2x + 5$$

↑ ↑ The slope, m, is 2. The y-intercept, b, is 5.

$$y = mx + b$$

Example: A line has a slope of 6 and a y-intercept of -3. Write the equation for the line.

The slope is 6, so m = 6. The y-intercept is -3, so b = -3.

Put those values into the slope-intercept form: $y = 6x - 3$

Example: Write the equation of a line that passes through points (3, 2) and (6, 4).

Only 2 things are needed to write the equation of a line: slope and y-intercept.

First, find the slope. $m = \frac{y_2 - y_1}{x_2 - x_1} = \frac{4-2}{6-3} = \frac{2}{3}$

Then, find the y-intercept. Choose either point. Let's use (6, 4). The x-value of this point is 6 and the y-value is 4. Put these values along with the slope into the equation and solve for b.

$y = mx + b$ $\quad 4 = \frac{2}{3}(6) + b$ $\quad 4 = \frac{12}{3} + b$ $\quad 4 = 4 + b$ $\quad 0 = b$

So the slope = $\frac{2}{3}$ and the y-intercept = 0. The equation of the line is $y = \frac{2}{3}x + 0$.

Help Pages

Solved Examples

Inequalities

An **inequality** is a statement that one quantity is different than another (usually larger or smaller). The symbols showing inequality are $<$, $>$, $\leq$, and $\geq$. (less than, greater than, less than or equal to, and greater than or equal to). An inequality is formed by placing one of the inequality symbols between two expressions. The solution of an inequality is the set of numbers that can be substituted for the variable to make the statement true.

A simple inequality is $x \leq 4$. The solution set, {..., 2, 3, 4}, includes all numbers that are either less than four or equal to four.

Some inequalities are solved using only addition or subtraction. The approach to solving them is similar to that used when solving equations. The goal is to get the variable alone on one side of the inequality and the numbers on the other side.

Examples: Solve $x - 4 < 8$.

$$\begin{array}{rcl} x - 4 & < & 8 \\ +4 & & +4 \\ \hline x & < & 12 \end{array}$$

1. To get the variable alone, add the opposite of the number that is with it to both sides.
2. Simplify both sides of the inequality.
3. Graph the solution on a number line. For $<$ and $>$, use an open circle; for $\leq$ and $\geq$, use a closed circle.

Solve $y + 3 \geq 10$.

$$\begin{array}{rcl} y + 3 & \geq & 10 \\ -3 & & -3 \\ \hline y & \geq & 7 \end{array}$$

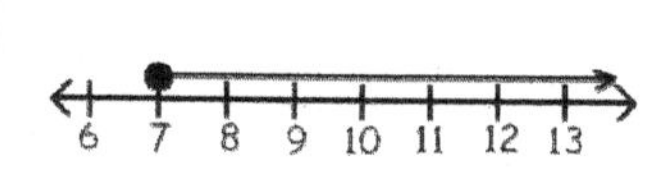

Some inequalities are solved using only multiplication or division. The approach to solving them is also similar to that used when solving equations. Here, too, the goal is to get the variable alone on one side of the inequality and the numbers on the other side.

The one difference that you must remember is this: If, when solving a problem you multiply or divide by a negative number, you must flip the inequality symbol.

Examples: Solve $8n < 56$.

$$\frac{8n}{8} < \frac{56}{8}$$

$$n < 7$$

1. Check to see if the variable is being multiplied or divided by a number.
2. Use the same number, but do the opposite operation on both sides.
3. Simplify both sides of the inequality.
4. Graph the solution on a number line. For $<$ and $>$, use an open circle; for $\leq$ and $\geq$, use a closed circle.

Solve $\frac{x}{-6} > 4$.

$$\frac{x}{-6} > 4$$

$$(-6)\frac{x}{-6} < 4(-6)$$

$$x < -24$$

Notice that during the 2nd step, when multiplying by -6, the sign "flipped" from greater than to less than.

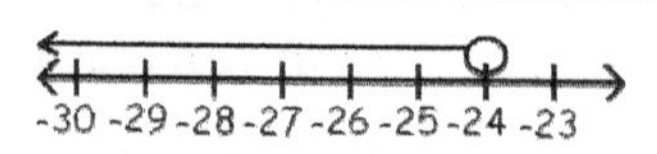

REMEMBER: When multiplying or dividing an inequality by a negative number, the inequality symbol must be flipped!

Help Pages

Solved Examples

Inequalities (continued)

Some inequalities must be solved using both addition/subtraction and multiplication/division. In these problems, the addition/subtraction is always done first.

Example: $2x - 6 \leq 6$

$$\begin{array}{rl} 2x - 6 \leq & 6 \\ +6 & +6 \\ \hline 2x \quad \leq & 12 \end{array}$$

$$\frac{2x}{2} \leq \frac{12}{2}$$

$$x \leq 6$$

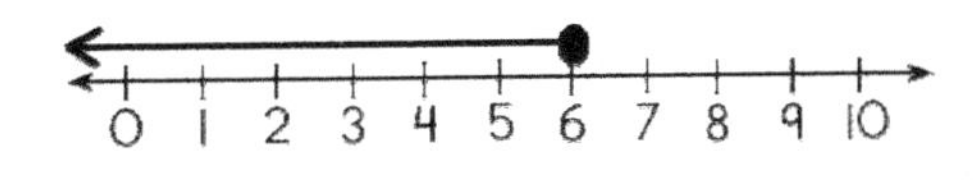

A compound inequality is a statement comparing one quantity (in the middle) with two other quantities (on either side).

$-2 < y < 1$ This can be read "y is greater than -2, but less than 1."

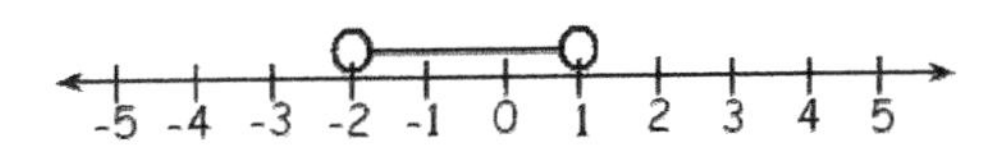

Integers

The rules for performing operations $(+, -, \times, \div)$ on integers are very important and must be memorized.

The Addition Rules for Integers:

When the signs are the same, add the numbers and keep the sign. When the signs are different, subtract the numbers and use the sign of the larger number.

$$\begin{array}{r} +33 \\ + +19 \\ \hline +52 \end{array} \qquad \begin{array}{r} -23 \\ + -19 \\ \hline -42 \end{array} \qquad \begin{array}{r} +35 \\ + -21 \\ \hline +14 \end{array} \qquad \begin{array}{r} -55 \\ + +27 \\ \hline -28 \end{array}$$

The Subtraction Rule for Integers:

Change the sign of the second number and add (follow the Addition Rule for Integers above).

$$\begin{array}{r} +56 \\ - -26 \\ \hline \end{array} \xrightarrow{\text{apply rule}} \begin{array}{r} +56 \\ + +26 \\ \hline +82 \end{array} \qquad \begin{array}{r} +48 \\ - +23 \\ \hline \end{array} \xrightarrow{\text{apply rule}} \begin{array}{r} +48 \\ + -23 \\ \hline +25 \end{array}$$

Notice that every subtraction problem becomes an addition problem, using this rule!

The Multiplication and Division Rules for Integers:

1. When the signs are the same, the answer is positive (+).

   $+7 \times +3 = +21$ $\quad -7 \times -3 = +21$

   $+18 \div +6 = +3$ $\quad -18 \div -6 = +3$

2. When the signs are different, the answer is negative (-).

   $+7 \times -3 = -21$ $\quad -7 \times +3 = -21$

   $-18 \div +6 = -3$ $\quad +18 \div -6 = -3$

+	×	+	=	+
-	×	-	=	+
+	×	-	=	-
-	×	+	=	-
+	÷	+	=	+
-	÷	-	=	+
+	÷	-	=	-
-	÷	+	=	-

Help Pages

Solved Examples

Matrix or Matrices

A **matrix** is a rectangular arrangement of numbers in rows and columns. Each number in a matrix is an element or entry. The plural of matrix is **matrices**.

The matrix to the right has 2 rows and 3 columns. It has 6 elements.

$$\begin{pmatrix} 0 & 4 & -1 \\ -3 & 2 & 5 \end{pmatrix}$$

In order to be added or subtracted, matrices must have the same number of rows and columns. If they don't have the same dimensions, they cannot be added or subtracted.

When **adding matrices**, simply add corresponding elements.

Example: $$\begin{pmatrix} 0 & 4 & -1 \\ -3 & 2 & 5 \end{pmatrix}+\begin{pmatrix} 2 & 1 & 3 \\ -2 & -6 & 4 \end{pmatrix}=\begin{pmatrix} (0+2) & (4+1) & (-1+3) \\ (-3+(-2)) & (2+(-6)) & (5+4) \end{pmatrix}=\begin{pmatrix} 2 & 5 & 2 \\ -5 & -4 & 9 \end{pmatrix}$$

When subtracting matrices, remember the subtraction rule for integers. A simple way to subtract matrices is to change the signs of every element of the second matrix. Then change the operation to addition and follow the rule for addition of integers (as shown in the previous example).

Example: $$\begin{pmatrix} -10 & 2 \\ 3 & -7 \end{pmatrix}-\begin{pmatrix} 5 & -3 \\ 6 & -1 \end{pmatrix}=$$ First, change all signs, then add.

$$\begin{pmatrix} -10 & 2 \\ 3 & -7 \end{pmatrix}+\begin{pmatrix} -5 & +3 \\ -6 & +1 \end{pmatrix}=\begin{pmatrix} (-10+(-5)) & (2+3) \\ (3+(-6)) & (-7+1) \end{pmatrix}=\begin{pmatrix} -15 & +5 \\ -3 & -6 \end{pmatrix}$$

Percent

Percent of change shows how much a quantity has increased or decreased from its original amount. When the new amount is greater than the original amount, the percent of change is called the **percent of increase**. When the new amount is less than the original amount, the percent of change is called the **percent of decrease**. Both of these are found in the same way. The difference between the new amount and the original amount is divided by the original amount. The result is multiplied by 100 to get the percent of change.

Formula: $$\text{\% of change} = \frac{\text{amount of increase or decrease}}{\text{original amount}} \times 100$$

Example: A sapling measured 23 inches tall when it was planted. Two years later the sapling was 36 inches tall. What was the percent of increase? Round the percent to the nearest whole number.

$$\left(\frac{36-23}{23}\right)\times 100 =$$

$$\left(\frac{13}{23}\right)\times 100 = 0.565$$ The sapling's height increased by 57% over the 2 years.

$$0.565 \times 100 = 57\%$$

Polynomials

A **Polynomial** is the sum of one or more monomials. Remember that a monomial is a number, a variable, or the product of numbers and variables with or without exponents. Here are some examples of monomials ($3x$, 7, $2xy$, $2y^2$, x^3) Polynomials may have any number of terms.

A **binomial** is a polynomial with exactly 2 terms. Examples: $3x - 7$, $2x + 5y$, $2y^2 + x^3$

A **trinomial** is a polynomial with exactly 3 terms. Examples: $3x - 7 + 2x$, $5y + 2y^2 - x^3$

Help Pages

Solved Examples

Polynomials (continued)

When **adding polynomials**, simply combine like terms. It is helpful to line up the like terms before adding.

Example: Add the following polynomials.

$$\begin{array}{r} 2x^3 - 6x^2 + x \phantom{{}-3} \\ + 3x^3 - 2x^2 \phantom{{}+x} - 3 \\ \hline 5x^3 - 8x^2 + x - 3 \end{array}$$

To subtract polynomials, use the subtraction rule for integers (change the sign of every term in the second polynomial and add).

Example: Subtract the following polynomials.

$$\begin{array}{r} 6n^2 - 4n + 5 \\ -\ 2n^2 + 2n - 3 \\ \hline \end{array} \Rightarrow \begin{array}{r} 6n^2 - 4n + 5 \\ + -2n^2 - 2n + 3 \\ \hline 4n^2 - 6n + 8 \end{array}$$

When **multiplying polynomials**, you must multiply each term in the first polynomial by each term in the second polynomial, and then combine like terms. (The process is similar to long multiplication.)

Example: Multiply. $(2b-2)(b^2+4b-5)$

$$\begin{array}{r} b^2 + 4b - 5 \\ \times \quad 2b - 2 \\ \hline -2b^2 - 8b + 10 \\ + 2b^3 + 8b^2 - 10b \phantom{{}+10} \\ \hline 2b^3 + 6b^2 - 18b + 10 \end{array}$$

When **multiplying two binomials**, the **FOIL method** can be used. In the FOIL method, the **F**irst terms in each binomial are multiplied, the **O**uter terms in each binomial are multiplied, the **I**nner terms in each binomial are multiplied, and then the **L**ast terms in each binomial are multiplied; finally all like terms are combined to get the product.

Example 1: Multiply using the FOIL method. $(3a+4)(a-2)$

The first terms are $3a$ and a; their product is $\mathbf{3a^2}$.

The outer terms are $3a$ and -2; their product is $\mathbf{-6a}$.

The inner terms are 4 and a; their product is $\mathbf{4a}$.

The last terms are 4 and -2; their product is **-8**.

The sum of these is $3a^2 + (-6a) + 4a + (-8)$. The product of the binomials is $\mathbf{3a^2 - 2a - 8}$.

Example 2: Multiply using the FOIL method. $(2x-1)(5x+3)$

The first terms are $2x$ and $5x$; their product is $\mathbf{10x^2}$.

The outer terms are $2x$ and 3; their product is $\mathbf{6x}$.

The inner terms are -1 and $5x$; their product is $\mathbf{-5x}$.

The last terms are -1 and 3; their product is **-3**.

The sum of these is $10x^2 + 6x + (-5x) + (-3)$.

The product of the binomials is a trinomial: $\mathbf{10x^2 + x - 3}$.

Help Pages

Solved Examples

Polynomials (continued)

The process of **factoring polynomials** is the reverse of multiplying them. When factoring, you are trying to find the simplest factors that can be multiplied together to get a polynomial.

Sometimes factoring is as easy as finding the largest common factor in each term and dividing all of the terms by that factor. You can think of this process as "undoing" the distributive property. When factoring any polynomial, you must first check to see if there is a common factor that can easily be divided out.

Example: Factor. $9x^2 + 6x = \frac{9x^2}{3x} + \frac{6x}{3x} = 3x(3x + 2)$

The largest factor that goes into both terms is $3x$. Divide each term by $3x$.
The result are the two prime factors: $3x$ and $3x + 2$

Once you've checked for any common factors, some trinomials can be factored further. The process of factoring a trinomial can be thought of as "undoing" the FOIL method. You are looking for 2 binomials that can be multiplied together to get the trinomial.

To understand this, look at this trinomial: $x^2 + 5x + 6$

This trinomial is the product of 2 binomials: $(x + 3)$ and $(x + 2)$. You need to understand where each term in the trinomial comes from.

The 1st term in the trinomial is the product of the 1st terms in the binomials: $x \cdot x = x^2$.

The last term in the trinomial is the product of the 2nd terms in the binomials: 3(2) = 6.

The middle term in the trinomial is the sum of the product of the inner terms $(3 \cdot x = 3x)$ and the product of the outer terms $(2 \cdot x = 2x)$ in the binomials: $3x + 2x = 5x$.

One way of representing this is $(a + b)(c + d) = ac + (bc + ad) + bd$.

Now, try to apply it to a different trinomial. Factor $x^2 + 11x + 18$.

$x^2 + 11x + 18$

The answer will be in this form:

$(\quad)(\quad) =$

x^2

$(x \quad)(x \quad) =$

18

Which one?

? $(x + 1)(x + 18) =$
Inner: 1x; Outer: 18x; Sum: 19x

? $(x + 2)(x + 9) =$
Inner: 2x; Outer: 9x; Sum: **11x**

? $(x + 3)(x + 6) =$
Inner: 3x; Outer: 6x; Sum: 9x

All of these are factors of 18. Which one results in the correct middle term?

1. Check to see if there is a factor common to all three terms. If so, factor it out. (There isn't.)
2. Look at the first term (x^2). What 2 terms can be multiplied to get x^2? $x \cdot x$ are the only factors. x will be the 1st term in both binomials.
3. Look at the last term (18). What 2 factors can be multiplied to get 18? There are several pairs of possibilities: 1 and 18, 2 and 9, 3 and 6. The 2nd terms in the binomials will come from these.
4. To choose the correct pair of factors, remember that each will be multiplied by the first terms and those products must add to the middle term.
5. The middle term is $11x$, so 2 and 9 must be the correct pair.

$(x + 2)(x + 9) = x^2 + 11x + 18$ ✓

Help Pages

Solved Examples

Polynomials (continued)

Example: What are the factors of $n^2 + 2n - 15$?

- The factors of n^2 are $n \cdot n$.
- The factors of -15 are 1 & -15, -1 & 15, 3 & -5, and -3 & 5.
- Each pair, multiplied by n, and added gives: $1n + -15n = -14n$, $-1n + 15n = 14n$, $3n + -5n = -2n$, $-3n + 5n =$ **$2n$**.

$(n \quad)(n \quad) =$

$(n - 3)(n + 5) =$

Since $2n$ is the correct middle term, the correct pair of factors is -3 and 5.

Some polynomials can be a bit more complicated to factor.
These are usually in the form of $ax^2 + bx + c$, where a is not 1. The process for factoring these polynomials is the same as those above, but there are many more possible factors to consider.

Example: Find the factors of $2x^2 - 7x + 3$.

1. The factors of $2x^2$ are $2x$ and x.
2. The factors of 3 are 1 & 3 and -1 & -3.

$(2x \quad)(x \quad) =$

Each combination of factors must be considered. Multiply the outer terms, then the inner terms and add them:

- Look at the combinations involving 1 & 3.
  $2x \cdot 1 = 2x$; $x \cdot 3 = 3x$; their sum is $5x$.
  $2x \cdot 3 = 6x$; $x \cdot 1 = 1x$; their sum is $7x$.
  Neither of these is correct.

The possible combinations are
$(2x + 3)(x + 1)$
or
$(2x + 1)(x + 3)$
Neither works.

- Look at the combinations involving -1 & -3.
  $2x \cdot -1 = -2x$; $x \cdot -3 = -3x$; their sum is $-5x$.
  $2x \cdot -3 = -6x$; $x \cdot -1 = -1x$; their sum is $-7x$.
  The last pairing is the correct one!

The possible combinations are
$(2x - 3)(x - 1)$
or
$(2x - 1)(x - 3)$
The last pair works.

The **difference of two squares** represents a special type of polynomial. This is a binomial where both terms are perfect squares, one term is positive, and one term is negative. When factoring a polynomial in this form $(a^2 - b^2)$, the results are always $(a + b)(a - b)$.

Example 1: What are the factors of $y^2 - 16$?

The first term, y^2, is a perfect square. y^2 — $(y + \quad)(y - \quad)$

The last term, 16, is a perfect square. $16 = 4^2$ — $(y + 4)(y - 4)$

Example 2: What are the factors of $49m^2 - 36$?

The first term, $49m^2$, is a perfect square. $(7m)^2$ — $(7m + \quad)(7m - \quad)$

The last term, 36, is a perfect square. $36 = 6^2$ — $(7m + 6)(7m - 6)$

Help Pages

Solved Examples

Probability

The **probability of two or more independent events** occurring together can be determined by multiplying the individual probabilities together. The product is called the **compound probability.**

P(A and B) = P(A) x P(B)

Example: What is the probability of rolling a 6 and then a 2 on two rolls of a die [P(6 and 2)]?

A) First, since there are 6 numbers on a die and only one of them is a 6, the probability of getting a 6 is $\frac{1}{6}$.

B) Since there are 6 numbers on a die and only one of them is a 2, the probability of getting a 2 is $\frac{1}{6}$.

So, P(6 and 2) = P(6) x P(2) = $\frac{1}{6} \times \frac{1}{6} = \frac{1}{36}$.

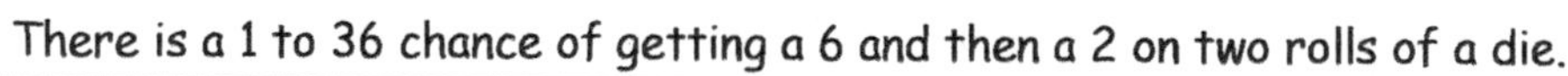

There is a 1 to 36 chance of getting a 6 and then a 2 on two rolls of a die.

Example: If you have a bag of 10 marbles, 4 green and 6 blue, what is the probability of picking a green marble and then a blue one without replacement?

A) First, the probability of picking a green marble is $\frac{4}{10}$.

B) Without replacing the first pick, there are now only 9 marbles in the bag, so the probability of picking a blue marble is $\frac{6}{9}$.

So, the probability is $\frac{6}{10} \times \frac{4}{9} = \frac{24}{90} = \frac{4}{15}$.

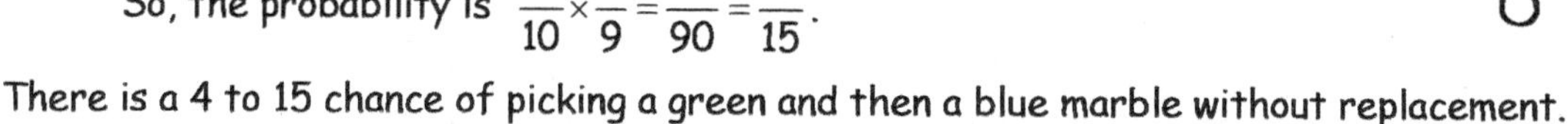

There is a 4 to 15 chance of picking a green and then a blue marble without replacement.

Radicals

A **radical expression** is an expression that contains a radical, such as a square root, cubed root, or other root. Radical expressions can be added, subtracted, multiplied, divided, and simplified much like rational expressions.

The square root of a number is equal to the product of the square root of its factor.

Example 1: $\sqrt{4x} = \sqrt{4} \cdot \sqrt{x} = 2\sqrt{x}$

Example 2: $\sqrt{32} = \sqrt{16 \cdot 2} = \sqrt{16} \cdot \sqrt{2} = 4\sqrt{2}$

Example 3: $\sqrt{9x^3} = \sqrt{9 \cdot x^2 \cdot x} = \sqrt{9} \cdot \sqrt{x^2} \cdot \sqrt{x} = 3x\sqrt{x}$

When **adding (subtracting) radical expressions**, the numbers under the radical must be the same or they can't be combined. This requires simplifying the radical first. Once the numbers under the radical are the same, the coefficients are added (subtracted) and the radical kept the same.

Example 1: Add. $3\sqrt{2} + 5\sqrt{2} = 8\sqrt{2}$

Example 2: Add. $5\sqrt{3} + \sqrt{48}$

$5\sqrt{3} + \sqrt{16 \cdot 3} =$

$5\sqrt{3} + \sqrt{16} \cdot \sqrt{3} =$

$5\sqrt{3} + 4\sqrt{3} =$

$9\sqrt{3}$

1. If the numbers under the radical are different, they must be simplified first.
2. Factor the numbers under the radical, looking for perfect squares.
3. If there are perfect squares, simplify.
4. Once the numbers beneath the radical are the same, add the coefficients.

Solved Examples

Radicals (continued)

Example 3: $3\sqrt{7} - 5\sqrt{14} + 2\sqrt{28}$

$$3\sqrt{7} - 5\sqrt{14} + 2\sqrt{28} =$$
$$3\sqrt{7} - 5\sqrt{7 \cdot 2} + 2\sqrt{4 \cdot 7} =$$
$$3\sqrt{7} - 5\sqrt{14} + 2\sqrt{4} \cdot \sqrt{7} =$$
$$3\sqrt{7} - 5\sqrt{14} + 4\sqrt{7} =$$
$$7\sqrt{7} - 5\sqrt{14}$$

Although $5\sqrt{14}$ can be broken down into factors, none of them is a perfect square, so $5\sqrt{14}$ is as simple as it can be.

Scientific Notation

Scientific notation is a shorthand method for representing numbers that are either very large or very small - numbers that have many zeroes and are tedious to write out.

For example, 5,000,000,000 and 0.000000023 have so many zeroes that it is not convenient to write them this way. Scientific notation removes the "placeholder" zeroes and represents them as powers of 10.

Numbers in scientific notation always have the form $c \times 10^n$ where $1 \leq c < 10$ and n is an integer.

Examples: $5{,}000{,}000{,}000 = 5 \times 10^9$ $\quad$ $0.000000023 = 2.3 \times 10^{-8}$

1. First locate the decimal point. Remember, if the decimal point isn't shown, it is after the last digit on the right.
2. Move the decimal point (either left or right) until the number is at least 1 and less than 10.
3. Count the number of places you moved the decimal point. This is the exponent.
4. If you moved the decimal to the right, the exponent will be negative; if you moved it to the left, the exponent will be positive.
5. Write the number times 10 to the power of the exponent that you found.

5,000,000,000

5.000,000,000.

5×10^9

The decimal point was moved 9 places to the left, so the exponent is +9.

0.000000023

0.00000002.3

2.3×10^{-8}

The decimal point was moved 8 places to the right, so the exponent is -8.

Who Knows???

Degrees in a right angle? .... (90°)

A straight angle? ....(180°)

Angle greater than 90°? .... (obtuse)

Less than 90°? ....(acute)

Sides in a quadrilateral? ....(4)

Sides in an octagon?.......................(8)

Sides in a hexagon? ....(6)

Sides in a pentagon? ....(5)

Sides in a heptagon? ....(7)

Sides in a nonagon? ....(9)

Sides in a decagon? ....(10)

Inches in a yard? .... (36)

Yards in a mile? ....(1,760)

Feet in a mile? ....(5,280)

Centimeters in a meter? .... (100)

Teaspoons in a tablespoon? ....(3)

Ounces in a pound? ....(16)

Pounds in a ton? ....(2,000)

Cups in a pint? ....(2)

Pints in a quart? ....(2)

Quarts in a gallon? ....(4)

Millimeters in a meter? ....(1,000)

Years in a century? .... (100)

Years in a decade? ....(10)

Celsius freezing? ....(0°C)

Celsius boiling? .... (100°C)

Fahrenheit freezing? ....(32°F)

Fahrenheit boiling? ....(212°F)

Number with only 2 factors? .... (prime)

Perimeter? ....(add the sides)

Area?..................................(length x width)

Volume?.............. (length x width x height)

Area of parallelogram? ..... (base x height)

Area of triangle?............($\frac{1}{2}$ base x height)

Area of trapezoid? ($\frac{base_1 + base_2}{2}$ x height)

Surface Area of a rectangular prism?.......
......................SA = 2(LW) + 2(WH) + 2(LH)

Volume of a cylinder?........................ ($\pi r^2 h$)

Area of a circle?..................................(πr^2)

Circumference of a circle? ....($d\pi$)

Triangle with no sides equal? ...... (scalene)

Triangle with 3 sides equal? .. (equilateral)

Triangle with 2 sides equal? ..... (isosceles)

Distance across the middle of a circle?
....(diameter)

Half of the diameter?..................... (radius)

Figures with the same size and shape?
.... (congruent)

Figures with same shape, different sizes?
.... (similar)

Number occurring most often?.......(mode)

Middle number? ....(median)

Answer in addition? ....(sum)

Answer in division?...................... (quotient)

Answer in multiplication? .... (product)

Answer in subtraction?..........(difference)